Basic Integrated Circuit Technology Reference Manual

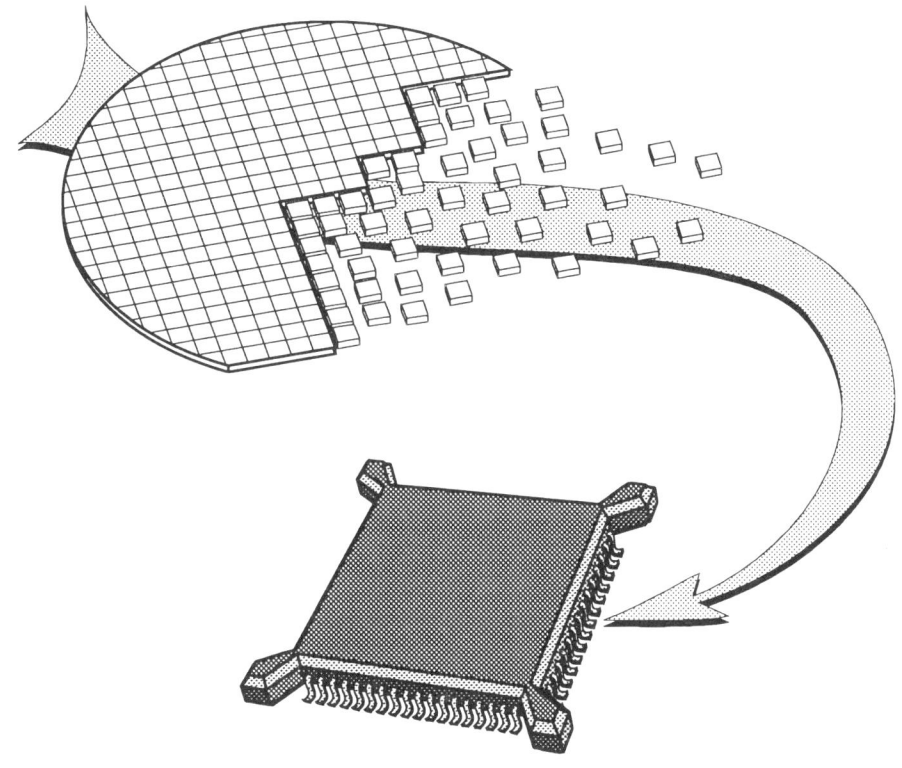

Editor:
Richard D. Skinner

ICE Staff Contributors:
Ron Bowman
Jim Griffin
Bill McClean

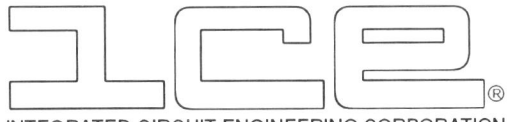

INTEGRATED CIRCUIT ENGINEERING CORPORATION
15022 N. 75th Street • Scottsdale, Arizona 85260-2476
Telephone: 602-998-9780 • Fax: 602-948-1925

ISBN 1-877750-24-7

Copyright © 1993 by Integrated Circuit Engineering Corporation

All rights reserved. No material contained in this report may be reproduced in whole or in part without the written permission of the publisher.

Printed on Recycled Paper

Basic Integrated Circuit Technology Reference Manual
Table of Contents

INTRODUCTION
- A. Vacuum Tubes I-1
- B. Transistors I-2
  - 1. Bipolar Transistors I-4
  - 2. MOS Transistors I-5
- C. Integrated Circuits I-6

SECTION 1. PRODUCTS
- A. Memory 1-1
  - 1. DRAMs 1-4
  - 2. PSRAMs 1-5
  - 3. SRAMs 1-5
    - a. Cache Memory 1-6
    - b. FIFO Memory 1-7
    - c. Multi-Port and VRAMs 1-8
  - 4. EPROM 1-9
  - 5. MROMs 1-10
  - 6. EEPROMs 1-10
  - 7. Flash 1-11
  - 8. Shadow and Battery-Backed SRAMs (NOVRAMs) 1-13
- B. Standard Logic 1-13
- C. Microprocessors and Microcontrollers 1-16
  - 1. Microcontrollers 1-16
  - 2. RISC vs. CISC Designs 1-17
- D. ASIC 1-17
  - 1. Custom 1-17
  - 2. Standard Cell 1-18
  - 3. Gate Array 1-18
  - 4. PLDs 1-20
    - a. PLAs, PALs 1-20
    - b. PGAs 1-21
- E. Analog 1-22
  - 1. Amplifiers 1-23
  - 2. Voltage Regulators 1-23
  - 3. Data Conversion 1-24
  - 4. DSP 1-26

Table of Contents

    5. Interface Circuits 1-26
    6. Analog Arrays 1-27
    7. Mixed-Mode 1-28
  F. System Terms 1-28
    1. LANs and Ethernet 1-28
    2. Displays - Active Matrix LCDs 1-29

SECTION 2. BASIC INTEGRATED CIRCUIT MANUFACTURING

  A. Starting Material 2-1
    1. Purification 2-1
    2. Czochralski Crystal Growing 2-2
    3. Sawing Crystals into Wafers 2-4
    4. Wafer Preparation 2-4
  B. Wafer Cleaning 2-5
  C. Dielectric Formation 2-9
    1. Thermally Grown Silicon Dioxide 2-9
      a. Atmospheric-Pressure Oxidation 2-10
      b. High-Pressure Oxidation 2-11
    2. CVD Silicon Dioxide 2-12
      a. Uses of CVD-Deposited Silicon Dioxide 2-12
      b. CVD Equipment 2-13
  D. Photolithography 2-15
    1. Overview 2-15
    2. Introduction 2-16
    3. Masks 2-18
      a. Reticles 2-19
      b. Photomasks 2-21
      c. E-Beam 2-21
      d. X-ray 2-21
    4. Photolithography Sequence 2-21
      a. Photoresist, Negative and Positive 2-21
        i. Exposure Wavelengths 2-22
        ii. Photoresist Parameters 2-24
      b. Wafer Preparation Before Photoresist Application 2-24
        i. Cleaning 2-24
        ii. Priming 2-24
      c. Photoresist Application 2-24
      d. Alignment/Exposure 2-25
        i. Contact Aligner 2-27
        ii. Proximity Printing 2-28
        iii. Projection Alignment 2-28
        iv. Direct Step on Wafer 2-29
          • Mix and Match 2-31
          • Pellicles 2-32

    v. E-Beam 2-33
    vi. X-Ray 2-35
     • X-Ray Sources 2-35
     • X-Ray Masks and Reticles 2-35
    vii. Summary 2-35
   e. Photoresist Developing 2-36
    i. Positive Resist 2-36
    ii. Negative Resist 2-38
    iii. Developing Methods 2-38
     • Batch Develop 2-38
     • Spray (Puddle) Develop 2-38
    iv. Plasma Descum 2-39
    v. Spin-Dry Process 2-39
    vi. Inspection After Develop 2-40
    vii. Post-Develop (Hard) Bake 2-40
   f. Etch 2-40
    i. Isotropic/Anisotropic 2-40
    ii. Dry vs. Wet Etch 2-42
    iii. Wet Etch 2-43
    iv. Dry Etch 2-44
   g. Photoresist Removal 2-44
E. Junction Formation 2-47
 1. Diffusion 2-47
  a. Predeposition 2-48
  b. Drive-In 2-50
  c. Process Control Measurements 2-51
   i. Resistivity 2-52
   ii. Junction Depth 2-54
 2. Ion Implantation 2-54
F. Epitaxial Deposition 2-59
G. Polysilicon Deposition 2-62
H. Metal Deposition 2-63
 1. Process Sequence 2-64
 2. Materials 2-66
 3. Methods 2-66
  a. E-Beam, Filament Evaporation 2-66
  b. Sputtering 2-68
 4. Electromigration 2-70
 5. Alloy (Sinter) 2-71
 6. Multilayer Interconnect 2-71
  a. Barrier Metals 2-72
  b. Vias 2-72
  c. Dielectric Isolation 2-72

I. Processes (NMOS, CMOS, Bipolar) 2-74
    1. CMOS 2-74
    2. NMOS 2-79
    3. Bipolar 2-81
    4. Bipolar (ECL) 2-82
  J. Assembly 2-83
    1. Die Separation 2-86
    2. Die Attach (Bond) 2-88
      a. Eutectic Attach 2-88
      b. Epoxy Attach 2-89
      c. Glass Frit Attach 2-89
    3. Wire Bond 2-89
      a. Thermocompression 2-90
      b. Ultrasonic 2-90
      c. Thermosonic Ball Bond 2-92
    4. Plastic Packages 2-98
      a. Molding 2-98
      b. Lead Trim and Form 2-100
    5. CERDIP 2-101
  K. Probe and Test 2-105
    1. Wafer Testing 2-105
      a. Test Patterns 2-105
      b. Probe Testing 2-106
    2. Final (Product) Testing 2-106
    3. Burn-In 2-106
  L. Summary of Semiconductor Manufacturing 2-107

3. PACKAGING
  A. Pad Pitch 3-1
  B. Dual In-Line Packages (DIPs) 3-2
  C. Single In-Line Packages (SIPs) 3-4
  D. Ceramic Flatpacks 3-5
  E. Pin Grid Arrays 3-5
  F. Surface Mount 3-6
    1. SOIC, QSOP 3-6
    2. LCC, PLCC, PQFP 3-6
    3. Thermal Coefficient of Expansion 3-8
    4. Flip Chip, Multi-Chip Modules (MCMs) 3-10
  G. Tape Automated Bonding (TAB) 3-12
  H. Chip-on-Board (COB) 3-12
  I. Silicon Substrates 3-13
  J. Ball Grid Array 3-15
  K. Special Packages 3-15

SECTION 4. TERMS BY MAJOR AREA OF INTEREST
   A. Tables 4-1
       1. Metric Unit Prefixes 4-1
       2. Length Conversion Factors 4-1
       3. Length, Area, Mass Conversion Factors 4-1
       4. Silicon Wafer Summary 4-1
       5. Electrical Units 4-1
       6. Periodic Table of Elements 4-1
   B. Key Semiconductor-Related Elements 4-1
   C. Key Semiconductor-Related Chemicals 4-2
       1. Acids 4-2
       2. Bases 4-2
       3. Solvents 4-2
       4. Gases – General 4-2
       5. Gases/Liquids – Process 4-3

Glossary G-1

List of Illustrations

I-1	Vacuum Tube	I-1
I-2	Gamma 3	I-2
I-3	Germanium Junction Transistor	I-3
I-4	Electronic Components	I-3
I-5	P-N Junction Formation	I-6
I-6	Diode Action	I-7
I-7	Bipolar Transistor – Planar Structure	I-7
I-8	Discrete and Integrated Transistors	I-8
I-9	N-Channel MOS Transistor	I-9
I-10	Enhancement-Mode MOS Transistor Operation (NMOS)	I-10
I-11	Simple Integrated Circuit	I-11
I-12	Semiconductor Technology Tree	I-12
1-1	System Partition of Personal Computer	1-1
1-2	Memory/Storage in a Computer	1-2
1-3	Types of Memory/Storage in a PC	1-2
1-4	Memory/Storage Alternatives	1-3
1-5	Memory/Storage in Various Machines	1-3
1-6	DRAM	1-4
1-7	SRAM 4T (Four-Transistor) Cell	1-5
1-8	SRAM 6T (Six-Transistor) Cell	1-6
1-9	Typical Memory System with Cache	1-7
1-10	DRAM Versus VRAM	1-8
1-11	Double-Poly Structure (EPROM/Flash Memory Cell)	1-9
1-12	EEPROM Cell Cross Section	1-10
1-13	Generic EEPROM Memory Cell	1-11
1-14	Flash EPROM Cell Connections	1-11
1-15	The NAND Versus NOR Contest	1-12
1-16	Block Diagram of the Xicor NOVRAM Family	1-14
1-17	Various Standard Logic Circuit Designs	1-15
1-18	Logic Families	1-16
1-19	Two-Input Multiplexer	1-18
1-20	The Basic Gate Array Cell	1-19
1-21	The Basic Gate Array Cell as a 2-Input NAND Gate	1-19
1-22	AND-OR PLA	1-20
1-23	Block Diagram of an AmPAL22XP10	1-20
1-24	PAL Block Diagram	1-21

List of Illustrations

Figure	Title	Page
1-25	Configurable Logic Block	1-22
1-26	Analog Example	1-23
1-27	Precision – Linear	1-24
1-28	Sampled Analog Waveshape	1-25
1-29	Real-World Signal Processing	1-25
1-30	Digital Signal Processing	1-26
1-31	Analog Array	1-27
1-32	Analog Array Component List	1-28
2-1	Polysilicon Creation	2-2
2-2	Types of Silicon Structures	2-2
2-3	Czochralski Silicon Crystal Grower	2-3
2-4	Inner-Diameter (ID) Sawing of Silicon Ingot	2-4
2-5	Wafer Preparation (for 150mm)	2-5
2-6	Sizes of Airborne Contaminants	2-6
2-7	Correspondence Between Scale of Integration (Design Rule) and Size of Particles to be Removed	2-7
2-8	Wafer Cleaning Techniques Prior to Thermal Processing	2-8
2-9	The Effects of SiO_2 Growth	2-10
2-10	Oxidation Systems	2-11
2-11	CVD Reactor Types	2-13
2-12	Block Diagram of a Low-Pressure Chemical Vapor Deposition System	2-14
2-13	Typical Reactions for CVD Depositions	2-14
2-14	Alignment and Exposure	2-15
2-15	Photolithography Using Positive Photoresist	2-16
2-16	The Layers Transferred to a Wafer During a Seven-Mask Process	2-17
2-17	Example of a Photomask	2-18
2-18	Die Placement in a Fixed Reticle of 21mm	2-18
2-19	The Design Process	2-19
2-20	Microlithography Roadmap	2-20
2-21	Photolithography Process Flow Chart	2-22
2-22	Characteristics of Negative and Positive Resists	2-23
2-23	Part of the Electromagnetic Spectrum	2-23
2-24	Photoresist Application	2-25
2-25	Wafer Patterning Systems	2-26
2-26	Contact Printer Optical Configuration	2-27
2-27	Dust Particle Scratching Photoresist and Photomask	2-27
2-28	Proximity Printer Optical Configuration	2-28
2-29	Principle of Scanning Projection Aligner	2-29
2-30	Direct Step on Wafer (DSW) Aligner Optical Configuration	2-30
2-31	Direct Step on Wafer (DSW)	2-31
2-32	Comparison of Stepper Reticle Sizes	2-32
2-33	Pellicle Protection Mechanism	2-33

List of Illustrations

Figure	Title	Page
2-34	E-Beam System	2-34
2-35	X-Ray Lithography	2-36
2-36	Comparison of Lithography Technologies	2-37
2-37	Photolithography Equipment Summary	2-37
2-38	Photoresist Develop	2-39
2-39	Photolithography Using Negative Photoresist	2-41
2-40	Major Visual Defect Examples	2-42
2-41	Vocabulary – Etch Process Section	2-43
2-42	Etch Methods	2-43
2-43	The Dry Etching Spectrum	2-45
2-44	Diagram of Barrel-Type Plasma Etcher	2-45
2-45	Planar Type Reactor	2-46
2-46	Selective Doping Using SiO_2 as a Mask	2-48
2-47	Profile of Dopant Present in a Wafer as a Function of Time	2-49
2-48	Diffusion Process Flow Chart for Predeposition	2-50
2-49	Diffusion Process Flow Chart for Redistribution	2-51
2-50	Oxidation/Diffusion Systems	2-52
2-51	Resistivity Measurement (ρ)	2-53
2-52	Illustration for Ohms per Square	2-53
2-53	Angle Lapping Measurement Method	2-54
2-54	Monochromatic Light Examinations	2-55
2-55	Silicon Gate MOS Cross-Section	2-55
2-56	Basic Concepts of Ion Implant	2-56
2-57	Configuration of a Typical 200KeV Ion Implanter	2-56
2-58	Path of an Implanted Ion	2-57
2-59	Step 1, The Adjustment of Thresholds	2-58
2-60	Step 2, Fabrication of the Depletion-Mode Transistors	2-58
2-61	Silicon Chemical Vapor Deposition Processes	2-59
2-62	Types of Deposited Films	2-60
2-63	Pictorial Representation of Epitaxial Growth	2-60
2-64	Epitaxial Growth	2-61
2-65	Low-Pressure Polysilicon Deposition	2-62
2-66	Block Diagram of a Low-Pressure Chemical Vapor Deposition System	2-63
2-67	Metallization Process Sequence	2-64
2-68	Metal Deposition Process Flow Chart	2-65
2-69	Metallization System	2-65
2-70	Conductor Material Properties Requirements for VLSI	2-66
2-71	Aluminum/Silicon Dissolution	2-66
2-72	Methods of Metal Deposition	2-67
2-73	Vacuum Evaporation System	2-67
2-74	Electron Beam Evaporation	2-68
2-75	Low-Pressure Sputtering	2-69

List of Illustrations

Figure	Title	Page
2-76	Aluminum Step Coverage Comparison	2-69
2-77	Planar DC Magnetron Sputtering	2-70
2-78	Electromigration Description	2-70
2-79	Three-Layer Metallization on IC Chip	2-71
2-80	Double-Layer Metal With Barriers	2-72
2-81	CVD Tungsten Thin Film	2-73
2-82	For Via Fill to Silicon	2-73
2-83	Twin-Well Silicon-Gate CMOS Manufacturing Process (1 of 3)	2-74
2-84	Twin-Well Silicon-Gate CMOS Manufacturing Process (2 of 3)	2-75
2-85	Twin-Well Silicon-Gate CMOS Manufacturing Process (3 of 3)	2-76
2-86	Twin-Well Silicon-Gate CMOS Process Flow (1 of 3)	2-77
2-87	Twin-Well Silicon-Gate CMOS Process Flow (2 of 3)	2-78
2-88	Twin-Well Silicon-Gate CMOS Process Flow (3 of 3)	2-79
2-89	NMOS Process	2-80
2-90	Bipolar Process Flow	2-81
2-91	Advanced Bipolar Digital Process Summary	2-82
2-92	IC Assembly Sequence	2-84
2-93	The Basis for Electronic Packaging	2-84
2-94	The Electronic Package	2-85
2-95	Package Design Factors	2-85
2-96	Assembly Processes	2-86
2-97	Wafer Saw	2-87
2-98	Examples of Wafer Saw Cut	2-87
2-99	Die Bonding	2-88
2-100	Die Attach Technology Comparison	2-89
2-101	Graphic Representation of a Thermocompression Wire Bond	2-90
2-102	Cross Section of Wedge Bond	2-91
2-103	Ultrasonic Wire Bonding	2-91
2-104	Ultrasonic Bonding Sequence	2-92
2-105	Ball Bonding	2-93
2-106	Thermocompression/Thermosonic Bonding Sequence	2-94
2-107	Inside Chamfer Discussion	2-95
2-108	Cross Section of Ball Bond	2-95
2-109	Loop Control of Wire Bonds	2-96
2-110	Wire Bond Failure Categories	2-96
2-111	Wire Bond Technology Comparison	2-97
2-112	Staggered Bonding Pads	2-97
2-113	Plastic Dual In-Line Package	2-98
2-114	Assembly Process Flow Chart for Plastic DIP	2-99
2-115	Lead Frames for Plastic DIP	2-100
2-116	Typical Lead Frame Magazine	2-101
2-117	Transfer Molding Procedure	2-102

List of Illustrations

Figure	Title	Page
2-118	Assembly Process Flow Chart for CERDIP	2-103
2-119	CERDIP Assembly Sequence	2-104
2-120	Bathtub Curve Prediction of Reliability	2-107
2-121	Integrated Circuit Manufacturing Process	2-108
3-1	IC Chip to PCB Lead Spacing	3-1
3-2	Plastic Dual In-Line Package	3-2
3-3	CERDIP Package	3-3
3-4	Side-Brazed Ceramic DIP	3-3
3-5	9-Pin Single In-Line Package	3-4
3-6	14-Pin Plastic QUIP	3-4
3-7	Multiwatt 11	3-5
3-8	Flatpack	3-5
3-9	Motorola VLSI Package with Die Cavity Below and Finned Heat Sink Above	3-6
3-10	Package Type Comparisons	3-7
3-11	QSOP Versus SOIC Packaging	3-7
3-12	Surface-Mount IC Packages	3-8
3-13	Surface Mount Leaded Connections	3-9
3-14	Top and Cross-Section Views of MM/PQFP Package	3-9
3-15	Properties of Package Insulator Materials	3-10
3-16	Flip-Chip Mounting	3-11
3-17	IBM MCM	3-11
3-18	Area Array TAB Concept	3-12
3-19	Hybrid Technologies Monolithic Chip-and-Wire	3-13
3-20	Chip-on-Board	3-13
3-21	Interconnection Substrate and IC Chips	3-14
3-22	Silicon-on-Silicon Packaging Concept	3-14
3-23	OMPAC Ball Grid Array from Motorola	3-15
3-24	X-Ray of Hitachi 4M DRAM ZIP Package	3-16
3-25	16M DRAM	3-16
4-1	Metric Unit Prefixes	4-4
4-2	Length Conversion Factors	4-5
4-3	Length, Area, Mass Relationships	4-5
4-4	Silicon Wafer Summary	4-6
4-5	Electrical Units	4-6
4-6	Periodic Table of the Elements	4-7

Foreword

This publication has been written to provide non-technical people who are interested in or associated with the integrated circuit (IC) industry a basic familiarity with IC products and manufacturing.

The publication is written as a reference manual rather than as a textbook. Where possible, explanations of terms and processes are initially described very simply, with increasingly detailed information following.

This publication cannot take the place of 20 years of IC industry experience. It does, however, give the IC industry newcomer an information base to draw upon to help facilitate communication and interaction with the IC industry veteran.

INTRODUCTION

The words, integrated circuits, semiconductor, microprocessor, and memory, are a part of the world we live in today. What is it all about and why is it important to you and me?

It's about the world of solid-state electronics and the revolutionary advances that have occurred during the second half of the twentieth century that have changed our lives and the way we do business.

A. VACUUM TUBES

At one time (for those of us who can remember the 40's and 50's), the term electronics was synonymous with vacuum tubes (Figure I-1). Tubes were used to produce electronic products like TV sets, radios, and computers. The first computers, like the GAMMA 3 shown in Figure I-2, were unbelievably large by today's standards and a lot of power was required to operate them.

Solid state technology in the form of integrated circuits (ICs) has become so overwhelming in the world today that the vacuum tube which started the electronics revolution has almost been forgotten. The IC has replaced the vacuum tube in the majority of the electronic applications but vacuum tube niches still continue to exist.

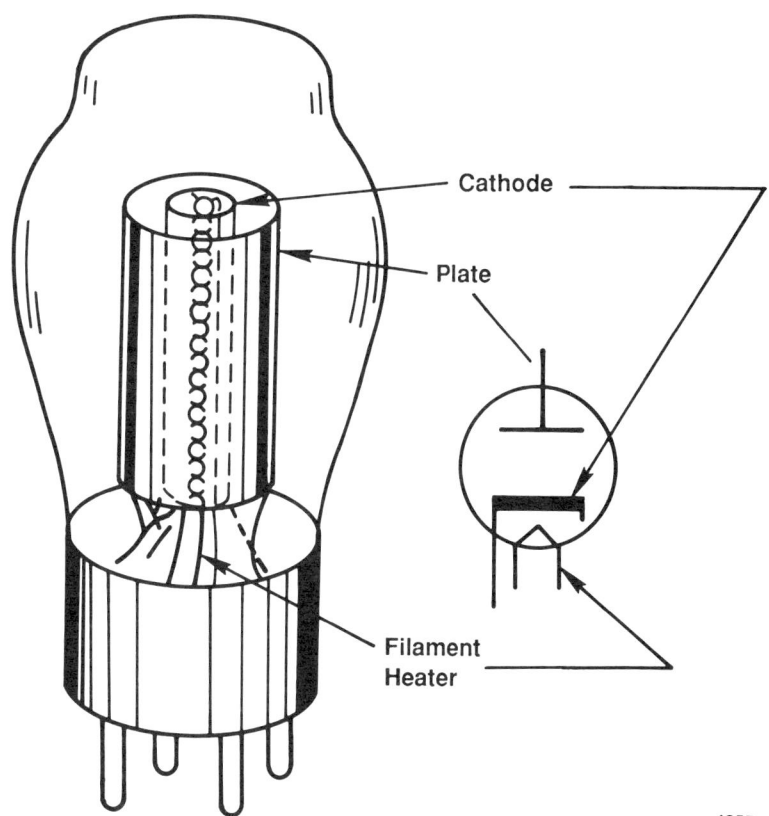

Figure I-1. Vacuum Tube

Introduction

Source: Illustrated Science & Invention Encyclopedia   15915

Figure I-2. GAMMA 3

As the name implies, the elements of the vacuum tube *must* be confined *inside* an evacuated container to function. Residing inside a vacuum is a necessary requirement just as the filament of an incandescent light bulb must be: the filament of either will oxidize in air because of the high operating temperature.

The vacuum tube functions as an electronic device by passing a large electrical current through the filament raising the filament temperature several hundred degrees. The heat liberated by the filament heats the cathode area by being placed close to but not touching the cathode. The heating of the cathode in the vacuum causes thermonic emission to occur. *(Thermonic emission means electrons can be "boiled off" the surface by heat).*

The plate in the vacuum tube is connected to a positive voltage and the cathode is connected to the negative side of the power supply. This difference in electrical potential causes the generated electrons to gain enough energy to be accelerated to the plate of the vacuum tube. After the electrons reach the plate they travel on through the external electrical circuit to the negative side of the power supply.

Another element can be added between the cathode and the plate called the grid. The voltage connected to the grid relative to the plate and cathode can control the movement of the electrons from the cathode to the plate. Thus, the grid element acts as a control valve.

Depending on the electronic requirement, several additional grid elements can be incorporated in this structure to further control the movement of electrons and provide very unique electrical responses.

The large filament currents needed for thermonic emission also causes large amounts of heat to be generated. This heat causes the vacuum tube to become very hot and the electronic system must liberate this heat either through ventilation or an added cooling system to maintain stability.

In March of 1948, Popular Mechanics magazine printed a rather bullish statement, "Where a calculator on the ENIAC is equipped with 18,000 vacuum tubes and weighs 30 tons, computers in the future may have only 1,000 vacuum tubes and perhaps weigh one and one half tons." The ENIAC, which was a contemporary of the GAMMA 3, was not nearly as powerful or fast as today's ordinary personal computer.

B. TRANSISTORS

The transistor (Figure I-3) was patented in 1947 by Bell Labs. This replacement for the vacuum tube was produced on a solid piece of semiconductor material. The term "semiconductor" originated from the Periodic Table of Elements (see Section 4), which uses the term to describe Family IVB (carbon, silicon, germanium, tin, and lead).

The first transistors were produced on germanium, but silicon became, and remains, the largest used material in the late 1960's.

The original Bell Labs research on the semiconductor was started on Group IVB elements in the Periodic Table. The elements in this section are carbon, silicon, germanium, tin, and lead. Based on the physical structure, tin and lead are conductors. This leaves carbon, silicon, and germanium as potential materials.

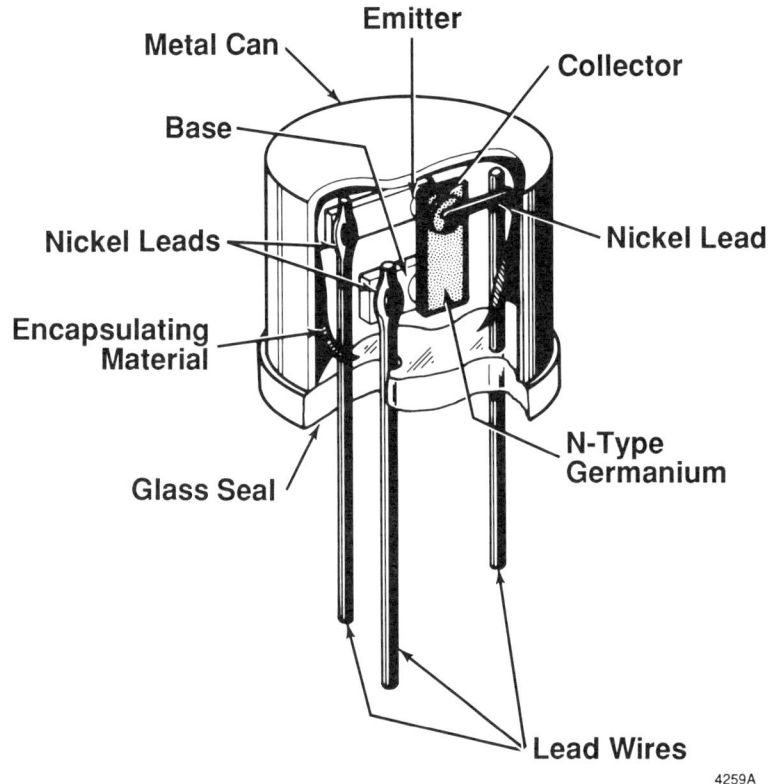

Figure I-3. Germanium Junction Transistor

Based on its hypothesis for semiconductor behavior, Bell Labs postponed research on carbon as an early materials choice. This left silicon and germanium. The Bell Labs research activities made faster progress in purifying germanium than silicon. As a consequence, early technology advanced germanium materials science to a production level by the early 1950's. The first production of germanium diodes and transistors was made at the Western Electric Works in Allentown, PA.

Research continued on silicon and by 1956 it was perfected to a sufficient level to start production. In 1958, both Fairchild Research Center and Bell Labs discovered that silicon would react with oxygen at elevated temperatures to form silicon dioxide (SiO_2). Germanium does not form a stable oxide. This has become a major factor in favor of silicon.

Additional research proved silicon had a larger band gap than germanium giving it a wider operating temperature range and lower junction leakage currents.

The previous two characteristics of a stable silicon dioxide and a higher operating temperature capability thrust silicon to the forefront in the early 1960's. By the late 1960's, silicon dominated semiconductor manufacturing and appears poised to be the material of choice for the remainder of the 90's.

Combinations of materials from Families IIIB and VB can also have semiconductor properties and be used to produce transistors. The largest used of the combinations is gallium arsenide. The interest in GaAs is due to the fact that electrons move more than five times as fast in GaAs as compared to silicon. This allows the transistors to operate very fast. However, as compared to silicon, GaAs manufacturing costs have always been too high to encourage widespread use of GaAs ICs.

Transistors, when used with other solid-state components (e.g., resistors, capacitors, and diodes — Figure I-4) allowed new electronic products that were lower power, smaller, faster, and more economical to be produced.

Figure I-4. Electronic Components

To date, solid state technology has not been successful in integrating inductors either within the silicon substrate or on top of the substrate. Inductors are generally connected externally to the chip/package or are attached in some hybrid form.

An inductor is a circuit element in which a change in current in the circuit element causes a change in the magnetic field which causes an induced counter electromotive force in the circuit.

A capacitor stores charge (electrons) as a voltage increases and looses charge as a voltage decreases.

Introduction

The capacitor circuit element can readily be formed on an IC. The capacitor can be formed by any combination of two conductors separated by a dielectric material.

A capacitor formed with an oxide insulating material has a constant capacitance value for a given dielectric constant, dielectric thickness, and area. Thus, silicon dioxide, silicon nitride, and polyimides are common dielectrics used in IC technology for constructing a capacitor.

A resistor is a circuit element that opposes the movement of charge through its body of material. Thus, the larger the resistance value the greater the opposition to charge movement and the smaller the amount of current passing through the resistor.

In IC technology, a resistor can be formed within the surface of silicon, from polysilicon, or by thin-film technology.

1. Bipolar Transistors

Before reviewing the details of a transistor, it is appropriate to first review the fundamental p-n junction called a diode. A diode cross section is shown in Figure I-5. A diode is formed by introducing into the silicon lattice dopant atoms that are the opposite type to the dopant already present. For example, in Figure I-5, if the original wafer was doped with boron to make the wafer P-type, then introducing phosphorous atoms into selected regions of the wafer surface would form a p-n junction. Thus a diode is a single p-n junction.

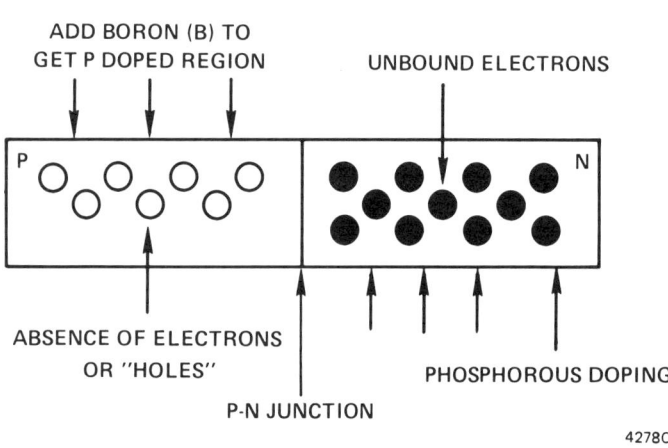

Figure I-5. P-N Junction Formation

Diode action is shown in Figure I-6. In this figure is illustrated the two possible electrical states for a diode: *(1) a forward biased junction causing the diode to conduct (2) a reverse biased junction does not allow conduction to occur.* Note the polarity of the battery in each state.

A transistor is two p-n junctions physically spaced very close together where the center region is common to the opposite doping on either side. A cross section of a typical n-p-n transistor is shown in Figure I-7.

The name of each region of the transistor is the emitter, base, and collector. The name of each region describes the function.

Introduction

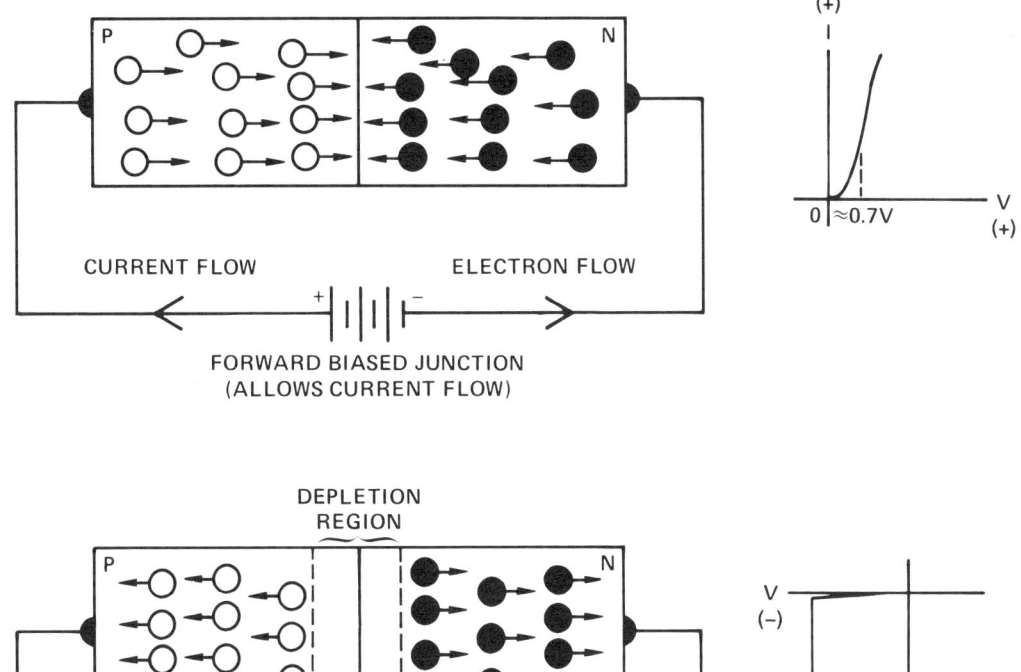

Figure I-6. Diode Action

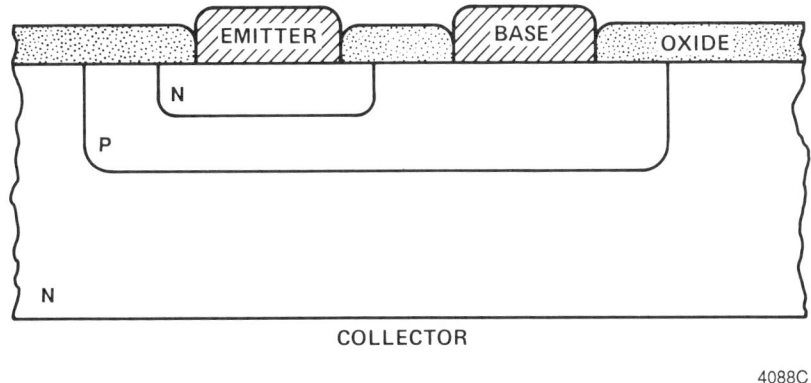

Figure I-7. Bipolar Transistor — Planar Structure

For a bipolar transistor to conduct and cause current to flow, the emitter-base p-n junction must be forward biased. The collector voltage (relative to the emitter reference) must be slightly higher than the voltage on the base. This causes the collector-base p-n junction to be reverse biased. Under these conditions, the forward bias on the emitter-base junction causes the emitter to emit electrons into the base region. The majority of these electrons have sufficient energy to travel through the base region *(the base is the reference or control area)* and get collected in the collector region wherein the electrons travel through the load to complete the electrical circuit.

As can be visualized from each regions function, the names of each region depicts the function. The name transistor was derived by the Bell Labs from the concept of moving charge (electrons) or transferring charge through varying degrees of resistance.

The transistor formed in a bipolar IC is a slight modification of the discrete transistor. The difference is illustrated in Figure I-8.

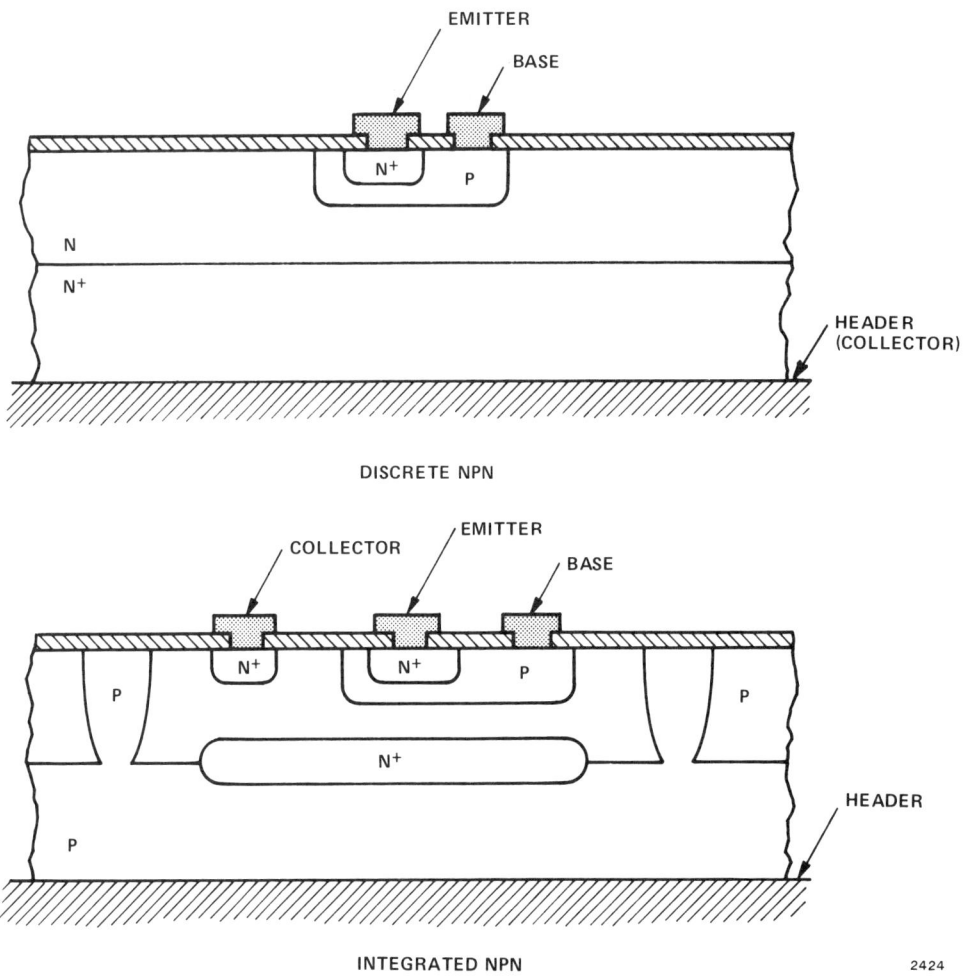

Figure I-8. Discrete and Integrated Transistors

Introduction

The transistor(s) used to build an IC must be electrically isolated from one another. Thus, a p-type substrate is used as part of the isolation structure and an n-type layer of silicon is added after the N+ buried layer is formed. This necessitates all the collector wiring be done on the top surface.

In a bipolar transistor the output (collector) current is equal to the base current multiplied by the current gain (h_{fe}) of the device. Small variations in base current (I_B) will cause large variations in collector current (I_C). In an amplifier circuit such as is in a radio, I_B is the input signal received by the radio and I_C is the larger signal driving the speaker(s). Bipolar transistors (and integrated circuits) are well suited for amplifier applications, which are linear or analog applications.

2. MOS Transistors

MOS transistors (and integrated circuits) are well suited for logic or digital applications. MOS stands for *M*etal (typically the "Gate" region of the transistor was aluminum) *O*xide (the insulating layer under the gate) *S*emiconductor (the silicon substrate being a semiconductor material).

The three parts to the MOS transistor are the source, drain, and gate. Functionally, the source is similar to the emitter of a bipolar transistor, the drain is similar to the collector, and the gate is similar to the base.

An N-channel MOS transistor that is not conducting current is shown in Figure I-9. With a positive voltage connected to the drain and the source connected to ground, the negatively charged electrons in the source have the inclination of moving across the channel into the drain and out the drain connection. The positively charged holes in the channel region, however, prevent the electrons from moving.

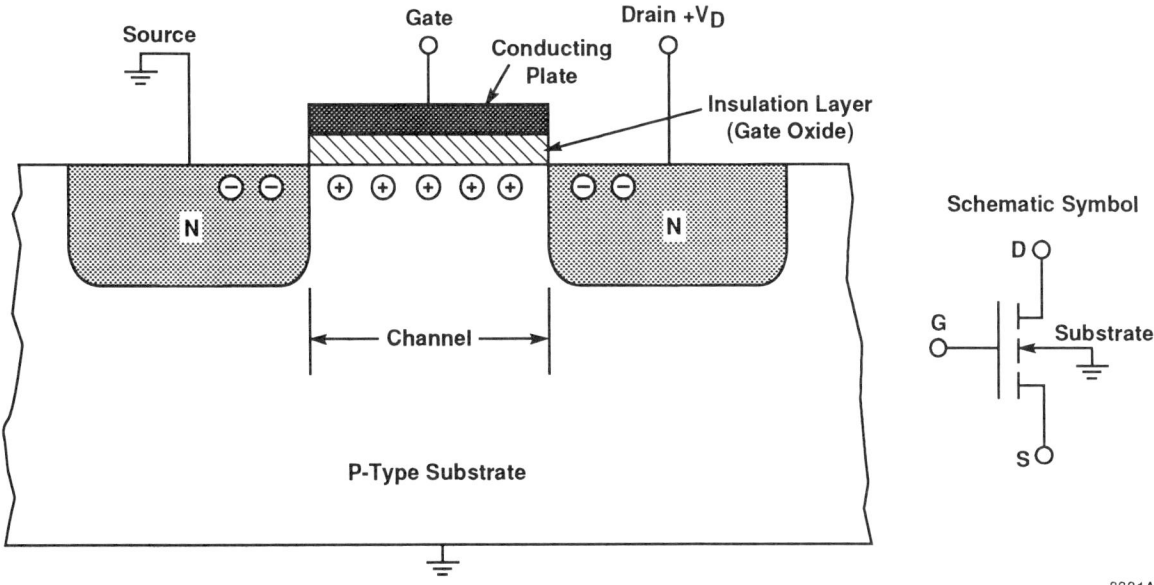

Figure I-9. N-Channel MOS Transistor

Introduction

The distance between the two N regions (i.e., channel) of this transistor is what is typically referred to as the feature size. Thus, when an IC producer refers to its one-micron IC process, one-micron is the distance between the source and drain. The shorter the distance, the faster the electrons can move across the channel to turn the transistor "on". Moreover, the smaller the channel region, the more transistors that can be put in a given area.

To turn the transistor on, the gate is connected to a positive voltage (Figure I-10). Although current cannot flow from the gate connection into the substrate because of the insulation layer (gate oxide), the positive voltage on the gate conductor pushes holes (identified by the + sign) away from the upper part of the channel. This then opens the way for electrons to move completely across the channel and out through the drain connection.

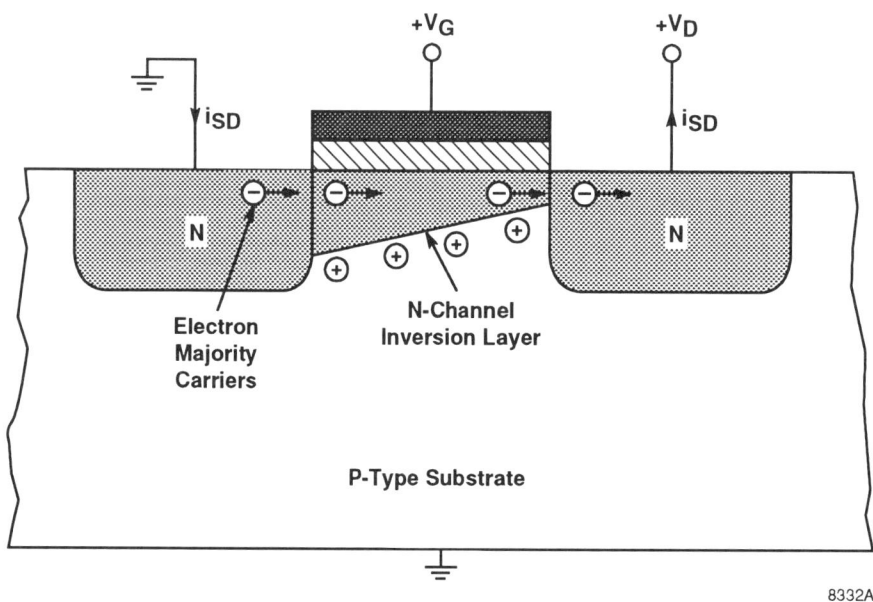

Figure I-10. Enhancement-Mode MOS Transistor Operation (NMOS)

The transition from off to on occurs within a few billionths of a second (nanoseconds) after the positive voltage is connected to the gate. The transistor is turned "off" when the positive voltage is removed from the gate. After removing the positive voltage the transistor goes back to the state it was in as shown in Figure I-9. Therefore, MOS transistors (and integrated circuits) are very well suited for on-off (digital) applications.

C. INTEGRATED CIRCUITS

The integrated circuit (Figure I-11) was developed in the late 1950's. Whereas previously only one transistor was produced on a single piece of semiconductor material, an integrated circuit has two or more transistors (up to tens of millions in the 1990's) and other solid state components (resistors, diodes, and capacitors) on a single piece of semiconductor material.

Introduction

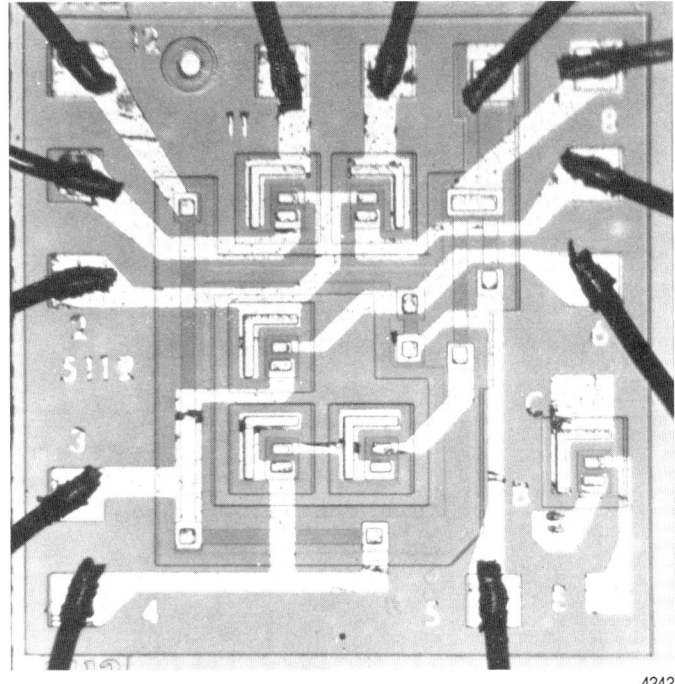

Figure I-11. Simple Integrated Circuit

There are two types of integrated circuit technologies, bipolar and MOS. Within each of these technologies are many subdivisions (Figure I-12).

Integrated circuits are also produced using a combination of bipolar and MOS transistors. They are called BiMOS or BiCMOS integrated circuits.

Over the history of the IC industry various processes have been developed to satisfy particular needs. Each of the IC processes have pros and cons associated with them. For example, the ECL bipolar process offers very fast transistor switching speeds but uses a lot of power and thus in comparison the MOS processes, cannot be used to produce high-density (e.g., ICs with millions of transistors) ICs. NMOS replaced PMOS IC technology in the 1970's because it offered much faster transistor switching speeds. CMOS displaced NMOS in the late 1980's in most applications because CMOS offers lower power consumption characteristics as compared to NMOS and can produce reliable ICs containing millions of transistors.

Introduction

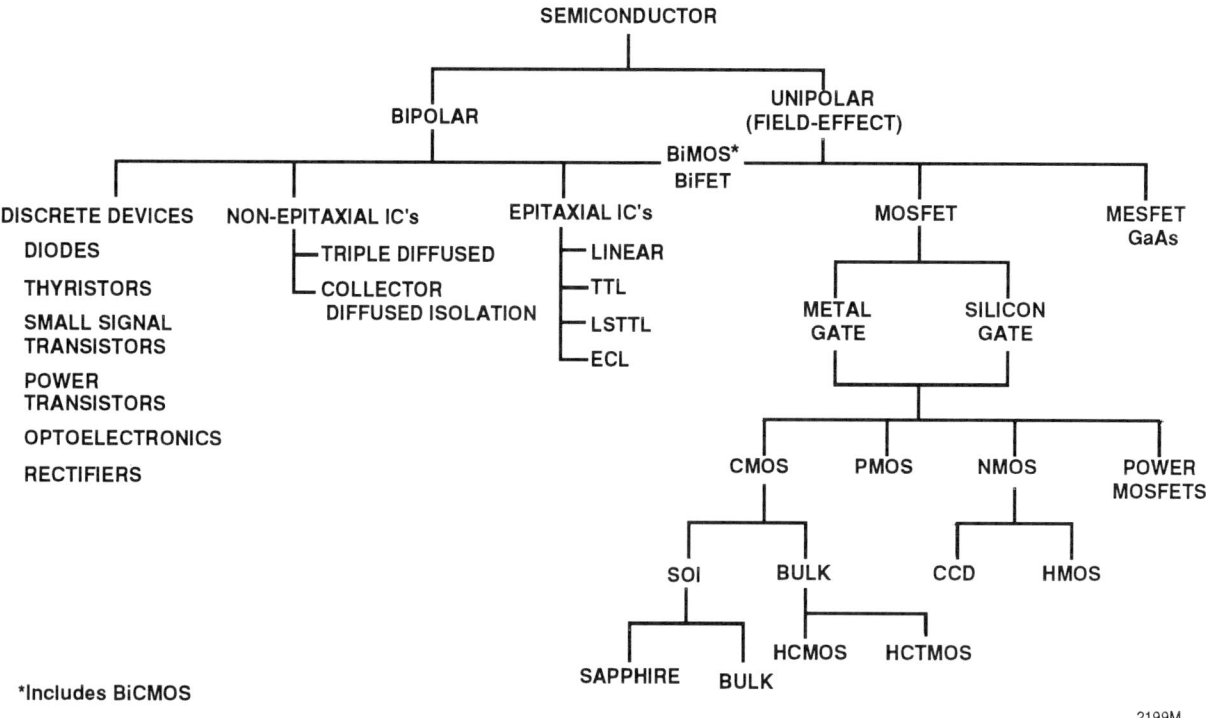

Figure I-12. Semiconductor Technology Tree

1 PRODUCTS

A. MEMORY

Memory and data storage take many forms, even in a single computer system. Figure 1-1 shows a simplified block diagram of a personal computer, indicating the various active sections needed to make the machine function. Several of the blocks are for storage of both the data and programs used by the computer. The CPU chip is the actual brain of the computer. But in order for this "brain" to function, it must call on data and programs stored in other parts of the computer. The CPU chip and the storage sections are shown in Figure 1-2.

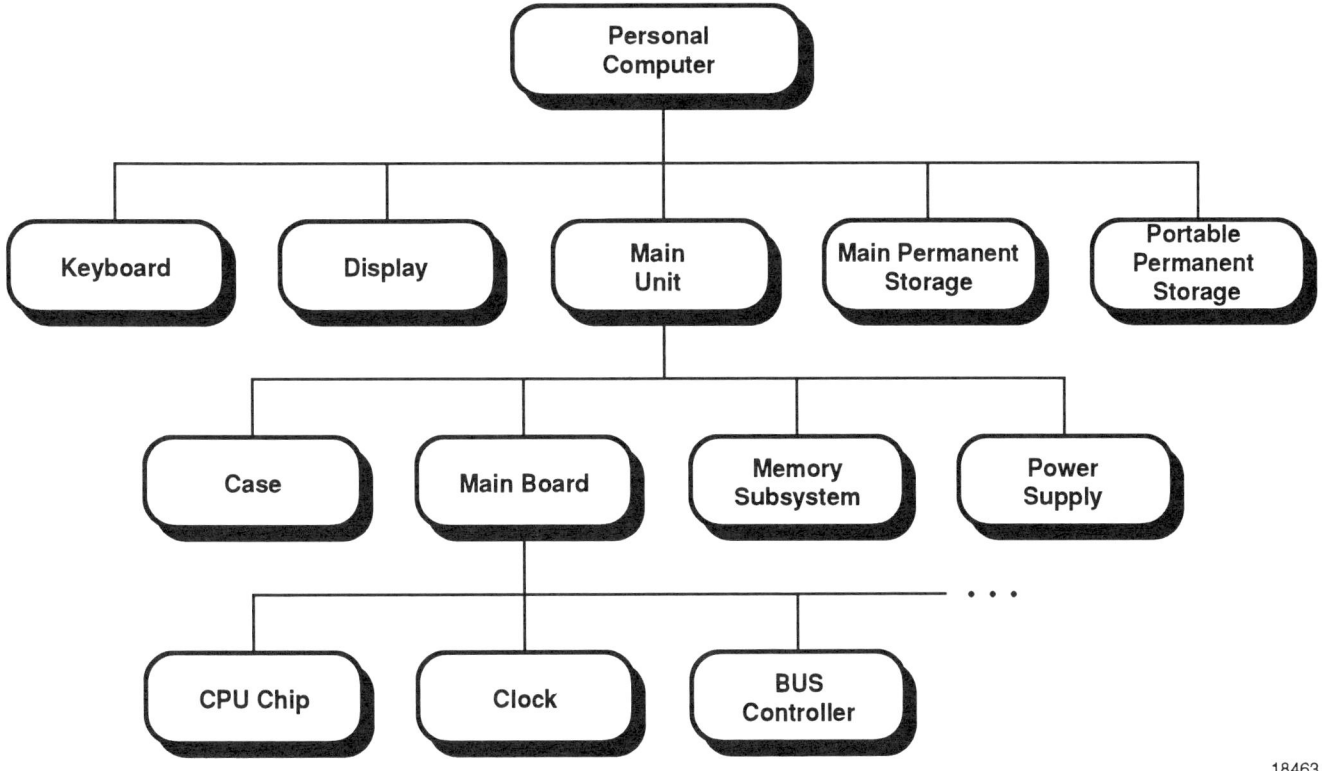

Figure 1-1. System Partition of Personal Computer

INTEGRATED CIRCUIT ENGINEERING CORPORATION

Products

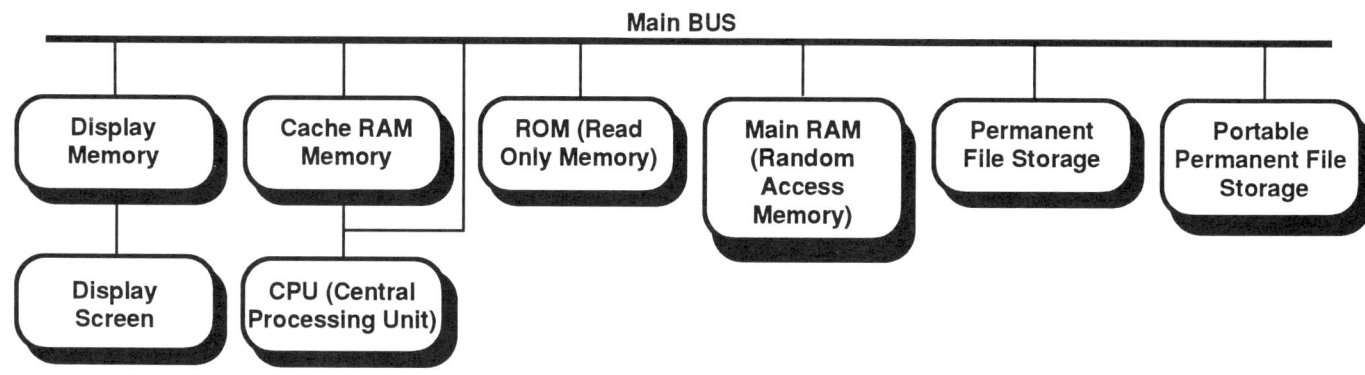

Figure 1-2. Memory/Storage in a Computer

To better understand the workings of a computer, the requirements of the blocks must be understood. Figure 1-3 describes the attributes of the various sections used for storage. Each function can be served by one or more types of semiconductor or magnetic device, as shown in Figure 1-4, which also describes the pros and cons of each alternative. As can be seen from this table, an alternative may be well suited for desktop machines (e.g., low cost, high power), but a different type is best suited for portable machines (e.g., low power). Mainframes, large minicomputers and large workstations use similar components, but these components are usually much larger in scale (and cost). Figure 1-5 shows a rough comparison of the differences in the computers described, without getting into performance and cost differences.

Circuit Block	Main Use	Typical Storage Size	Driving Attribute	Other Attributes
Main permanent file storage	Active data storage	50 MBytes or more	Low cost	Non-volatile and alterable
Portable permanent file storage	File exchange and backup	0.5 to 3 MBytes	Removable	Non-volatile and alterable
ROM	Machine startup and screen drawing routines	0.2 to 2 MBytes	Non-volatile	Reasonably fast, non-alterable
Main RAM	General temporary storage during operation	8 MBytes	Reasonably fast (<100nsec)	Low cost and alterable
Cache RAM	Very fast temporary storage	4 to 256 KBytes	Very fast (<15nsec)	Alterable
Video RAM	Stores data displayed on screen	1 to 24 bits/pixel	Fast enough for screen refresh	Low cost and dual access for fast read/write

Figure 1-3. Types of Memory/Storage in a PC

Requirement	Storage Alternative	Advantages	Disadvantages
Main permanent file storage	Hard Disk	Low cost	Large size, high power, not rugged
	Flash EEPROM	Rugged, small, light	High cost
Portable permanent file storage	Floppy disks	Very low cost media	Small storage per disk
	Flash EEPROM	Rugged, small, light	High cost
	Optical & Magneto-optical	Very large capacity, low cost media	High entry cost, not a standard yet
ROM	Mask ROM	Very low cost	Un-alterable
	EPROM	User alterable at factory	Slow to program
	EEPROM	Field alterable	Very high cost
	Flash EEPROM	Field alterable	High cost
Main RAM	DRAMs	Lowest cost	High power
	SRAMs	Low power, fast	High cost
	PSRAMs	Low power	Medium cost
Cache RAM	SRAMs	Fast	Very high cost
Video RAM	DRAMs	Low cost	Single mode access
	VRAMs	Low system cost, fast	Expensive for small systems

Figure 1-4. Memory/Storage Alternatives

	Size Main Memory	Size of Cache Memory	Permanent File Storage Size	Portable Permanent File Storage
Laptop	6 Meg	0	80 MByte	Floppy
PC	8 Meg	32K	200 MByte	Floppy
Workstation	32 Meg	256K	300 MByte	Floppy
Workstation as Network server	128 Meg	512K	1 GByte	Tape
Super Mini	256 Meg	512K	40 GByte	Tape
Mainframe/ Super Computer	1 GByte	512K each Processor	100 GByte	Tape

Figure 1-5. Memory/Storage in Various Machines

Products

The semiconductor devices in Figure 1-4 are described below in terms of both architecture and storage mechanism. Each type also requires slightly different manufacturing steps, leading to very different wafer costs.

1. DRAMs

DRAM (Dynamic Random Access Memory - Figure 1-6). This is the main memory used for all desktop and larger computers. Each "storage cell" consists of a single MOS transistor and storage capacitor, and can store one bit of information. The DRAM is unique in the fact that the data in each "cell" is lost over time due to leakage of the cell transistor, and must be refreshed several times each second. This refresh is done automatically and is actually not a problem in desktop computers.

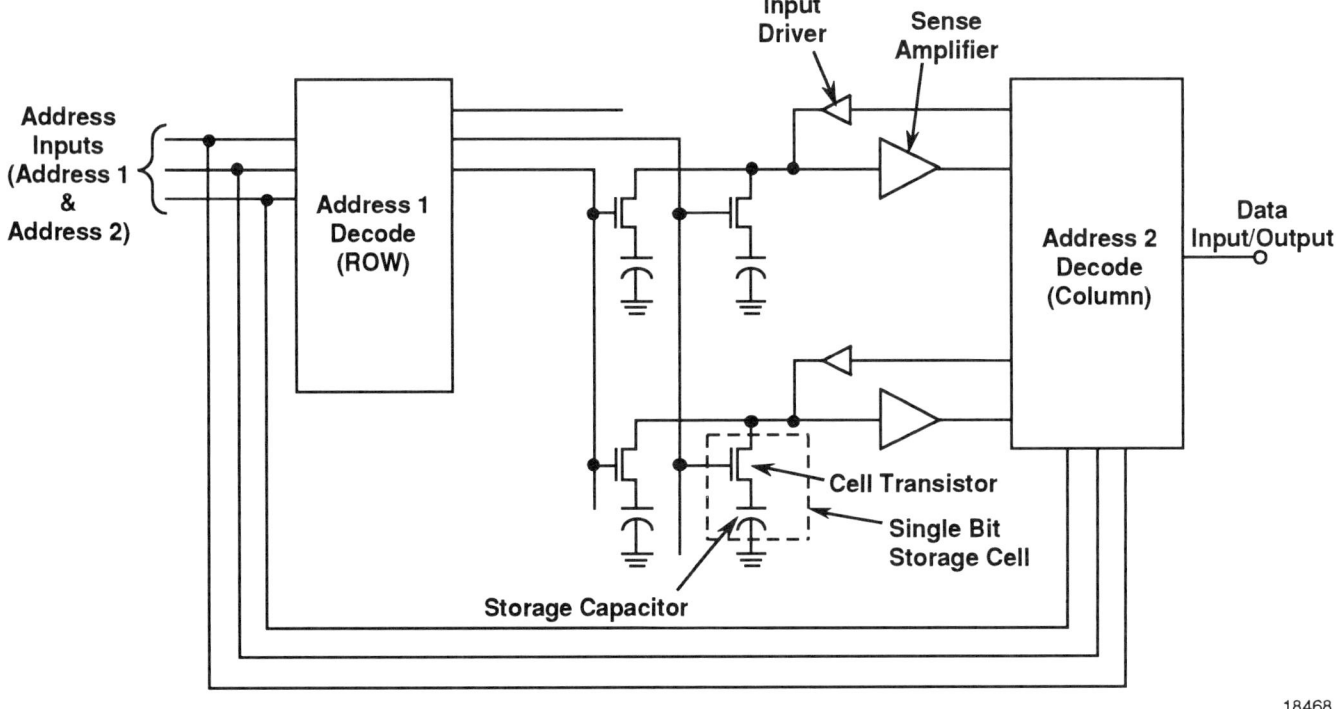

Figure 1-6. DRAM

The big advantage of the DRAM is that the amount of silicon required per bit of storage is very small, and the cost per bit of storage is the lowest of all current memories. DRAMs do suffer from a speed problem, as the circuitry on the chip required to read the data in each cell is inherently slow. As such, the DRAM speeds have not been able to improve at the same pace as the speeds of the CPUs.

The early CPUs were introduced with clock speeds of 1MHz (one million cycles per second). Today the faster CPUs in desktop PCs are 100MHz, a 100 times improvement. The early DRAMs had access cycle times (the time required for the DRAM chip to supply the data back to the CPU) of 250nsec, and the fastest units today are 50nsec or less, an improvement of more than five times.

Newer technologies are now being introduced to reduce the speed bottleneck. One innovation was to synchronize the read and write commands to the DRAM to the system clock, thereby enabling the CPU chip to receive data on a more rapid basis once the first address instruction has been received by the DRAM memory. These newer chips are called synchronous DRAMS for this reason.

2. PSRAMs

PSRAMs (Pseudo Static Random Access Memory) are actually DRAMs optimized for very low power consumption and designed with a simpler interface to the system bus. The cell structure is nearly identical to the DRAM, so the cost is low. The PSRAM sacrifices speed for the lower power consumption, so the main application is laptop computers where a small sacrifice in speed is preferable to reduced battery life. The PSRAM is a good compromise between the DRAM and the more expensive SRAM.

3. SRAMs

SRAMs (Static Random Access Memory) are designed to fill two needs. The first is the need for memory that can operate at the speed of the CPU. In this role it is called cache memory, which will be discussed later. These chips may be accessed in as little as a few nanoseconds, versus 50 to 100 nanoseconds for a DRAM. In the case of most high-end processors a large block of SRAM is integrated onto the processor chip.

The fast SRAM cell is usually composed of four transistors and two resistors. The storage cell area of an SRAM is about four times as large as the cell of the comparable generation DRAM. The data in an SRAM cell is volatile, i.e., the data is lost when the power is removed. However, the data does not "leak away" like in a DRAM. The SRAM doesn't require the refresh cycle and the associated power consumption. Figure 1-7 shows the cell schematic of the four-transistor cell (4T cell) SRAM.

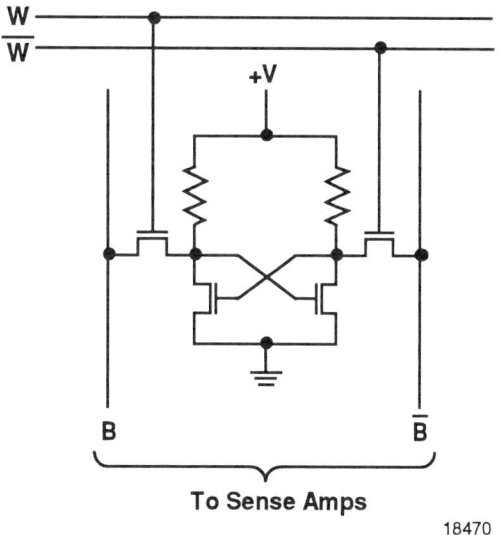

Figure 1-7. SRAM 4T (Four-Transistor) Cell

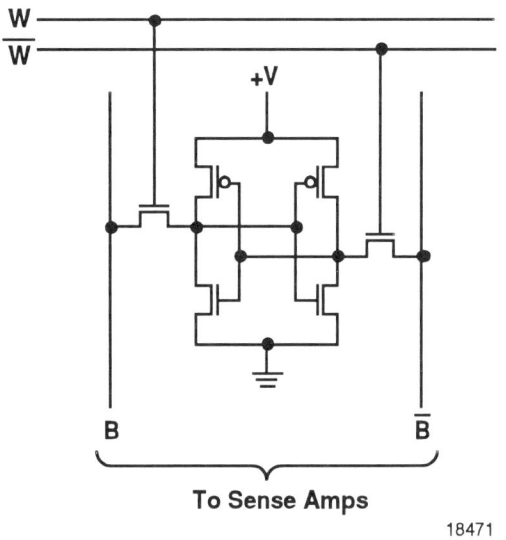

Figure 1-8. SRAM 6T (Six-Transistor) Cell

A second version of the SRAM is designed to accentuate the low-power capability while sacrificing speed and cost. These SRAMs are composed of six-transistor cells. This structure is shown in Figure 1-8. These devices are used where extremely low power consumption is mandatory, such as palm top computers operating from AA batteries.

Self-timed SRAMs are a special case for very high-performance applications. The address information is passed from the CPU to the SRAM in a short burst, instead of through the entire read or write cycle as with the standard SRAM. This allows the system designer more latitude in the design of the timing of read/write cycles, as the other timing signals are created inside the SRAM chip and presented a fixed period of time later to the CPU. Unfortunately, these features add complexity and cost to the SRAM design. It is usually only found in very high speed mainframe computers.

a. Cache Memory

To implement a cache memory requires the use of special circuits that keep track of which data is in both the main memory (DRAM) and the SRAM cache memory, and which data is only in the main memory. This function acts like a directory that tells the CPU what is or is not in cache.

The function can be designed with standard logic components, with SRAM chips for the data storage. An alternative is the use of special memory chips called Cache Tag RAMs that, in conjunction with the SRAM data memory chips, perform the entire function. Figure 1-9 shows both the Cache Tag RAM and the Cache Buffer RAM along with the main memory and the CPU (processor).

Another innovation is the CDRAM, a DRAM with a cache controller and cache memory (made with SRAM) integrated on each DRAM chip. This approach effectively offers a much simpler solution for the systems designers, and improved performance over a system with no SRAM cache memory.

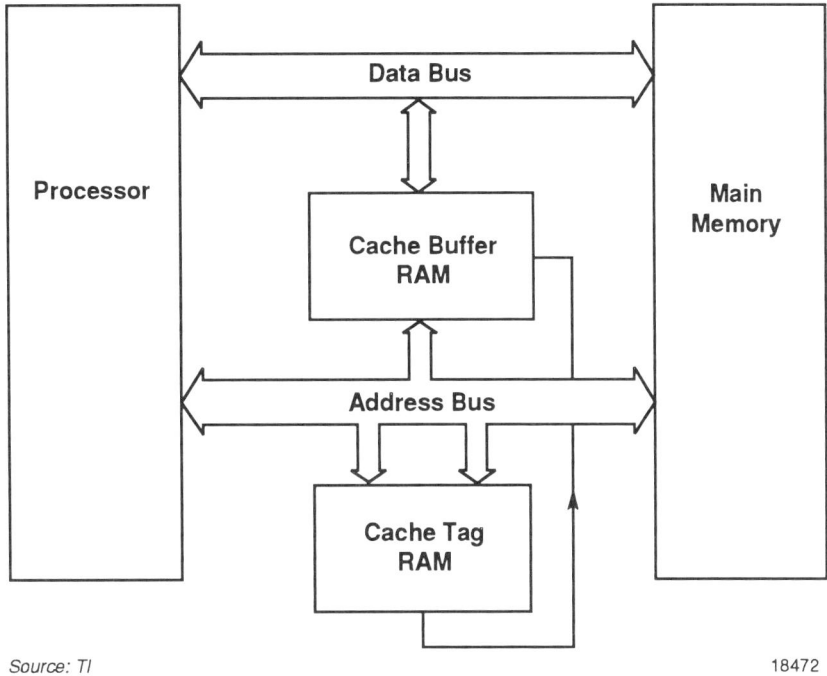

Figure 1-9. Typical Memory System with Cache

b. FIFO Memory

A FIFO (First In, First Out) memory is a specialized memory used for temporary storage to aid the timing of non-synchronized events. A good example of this is the interface of a computer system to a Local Area Network (discussed later).

Often, a computer system will receive data along the network at some inopportune time, such as when the CPU is in the middle of calculating the cells of a spread sheet or saving a file to disk. In these cases, there is a priority conflict, because the data coming along the network must be captured immediately or it will be lost.

One approach is for the computer to "drop what its doing" and service the network request. This approach is quite inefficient, but can be done in software as part of the operating system. This is a relatively inexpensive approach.

Another alternative is to store the data received from the network until the current task is complete. This can be done with special circuitry that stores the network data in special memory chips as it is received. This memory area is operated on a "First In, First Out" basis, because that is the way the data is received.

Products

The FIFO memory can be constructed with logic components and SRAM chips, or with the use of FIFO memory devices. The latter case is usually much more efficient use of board space and is typically found in all high-speed communications applications.

c. Multi-Port and VRAMs

Multi-port (usually two port, but sometimes four port) memories are specially designed chips using either SRAM or DRAM memory cells, but with special on-chip circuitry that allows multiple ports (paths) to access the same data, at almost the same time.

One application is where multiple CPUs are tied together by common memory. The multi-port chip acts as a bridge for temporary storage between two CPUs, so that each can operate at maximum efficiency even when data from the other is needed. Because of the speed requirements, only SRAMs can be used for this.

A special case of the two-port memory is the Video Random Access Memory (VRAM). The VRAM is used for high-performance video memory, usually for large screens (high pixel count), where the screen information storage is very large. This data storage area must be "read from" and "written to" very rapidly. In small-screen computer applications, standard DRAMs are often used.

The VRAM is usually constructed using the same fabrication process and cell structure as the DRAM. The VRAM, however, has special circuitry for multiple access to the memory data — one for writing and another for reading. Figure 1-10 shows the extra size required for this circuitry. Both chips have one million bits of storage, and are constructed with the same storage cell size.

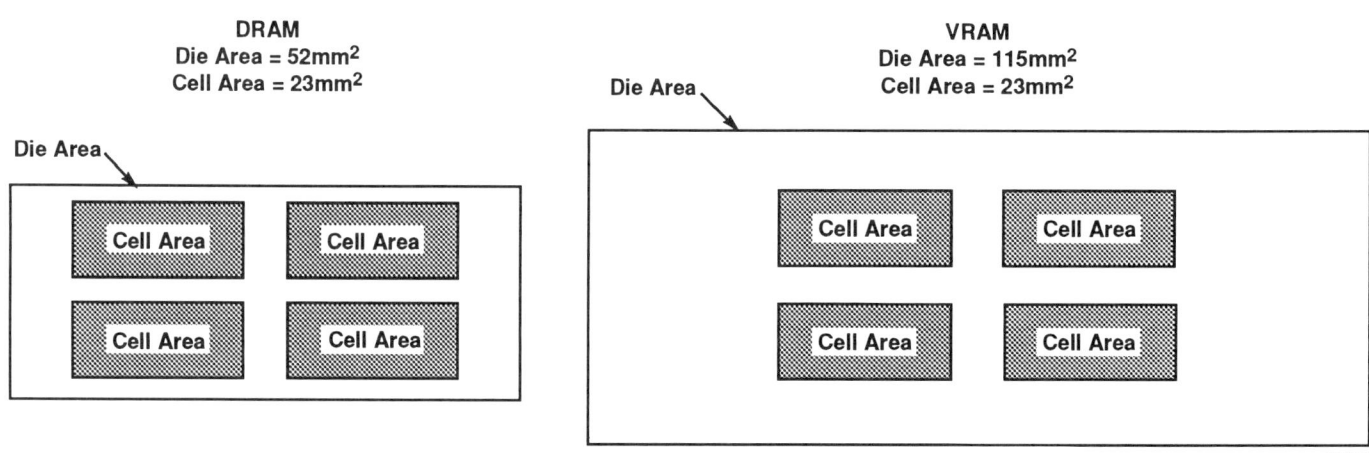

Figure 1-10. DRAM Versus VRAM

4. EPROM

EPROM (Electrically Programmable Read Only Memory) is a special case of semiconductor memory that is truly non-volatile, i.e., the memory retains the stored information when the power is removed. The device is programmed by forcing an electrical charge on a small piece of polysilicon material (called a floating storage gate) located in the memory cell. When this charge is present, the cell is "programmed." The programming (write) cycle takes several hundred milliseconds per byte. The read time is comparable to that of fast DRAMs. Figure 1-11 shows the cell used in a typical EPROM. The floating gate is where the electrical charge is stored.

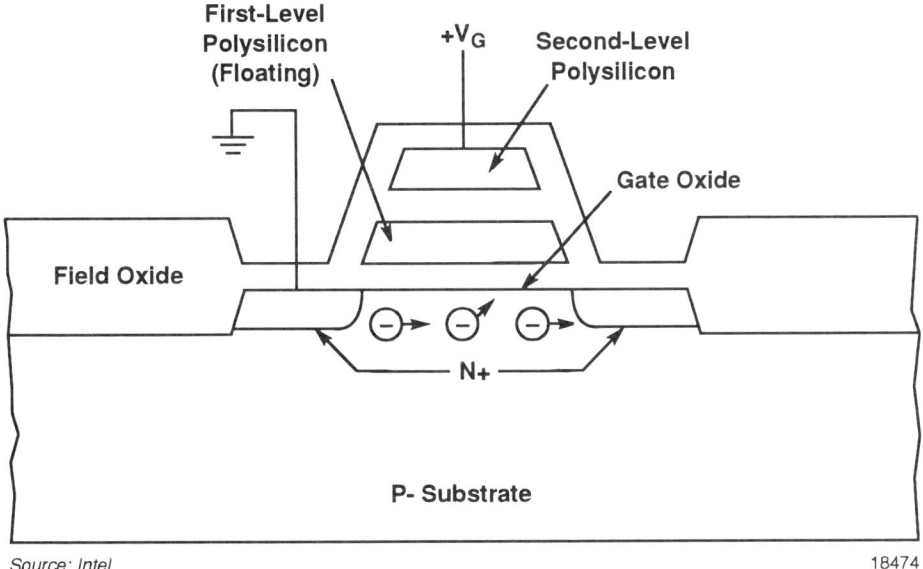

Figure 1-11. Double-Poly Structure (EPROM/Flash Memory Cell)

Some EPROMs are packaged with glass windows over the chips. If a strong ultraviolet light is shined on the chip for several minutes, the entire chip can be erased. Another alternative is the One-Time Programmable (OTP) version packaged in standard black opaque plastic. In this case, the chip can only be programmed once, hence the name.

The erasure capability is the EPROM's main advantage over the Mask Programmed ROM. It allows the user to buy mass-produced devices off the shelf and program each device for a specific need. The Mask Programmed ROM, however, is the lowest cost ROM available.

5. MROMs

MROMs (Mask Programmable ROMs) are used in certain applications where the identical "program" is repeated in each of hundreds of thousands of sockets. Examples of this are electronic games, the special screen graphics routines in the Macintosh computer, and the very small startup circuit in PCs (required to read the disk drive before the operating system is loaded), called BIOS. MROMs come from the IC factory already programmed and cannot be changed. This limits their applications to very high volume and stable products.

6. EEPROMs

The EEPROM (Electrically Erasable Programmable Read Only Memory) is actually incorrectly named, as it is really not a "Read Only Memory" at all, but can be electrically written as well as read. The cell structure is similar to the EPROM, but the storage transistors are manufactured so that, when the high voltages used for programming are applied in the reverse direction, the program will be erased. The programming and erasure mechanisms are both relatively slow, tens of milliseconds. The read times are as fast as DRAMs. Figure 1-12 shows the cross section of an EEPROM storage transistor.

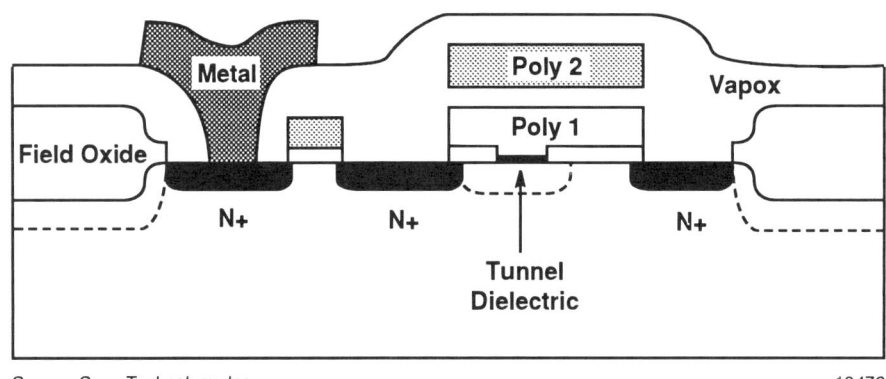

Source: Seeq Technology, Inc.   18476

Figure 1-12. EEPROM Cell Cross Section

The advantage of this device is the in-circuit erasure capability. The disadvantage is the very large cell size required to allow both the high positive and negative voltages (for writing and erasing, respectively) to be directed to each individual cell. This is shown in Figure 1-13.

Another disadvantage of EEPROMs is the limitation on the number of erase cycles allowed. This limitation is due to the slight damage created in the oxide film each time the device is erased. Typically, the limit is 100,000 erase cycles per cell.

Products

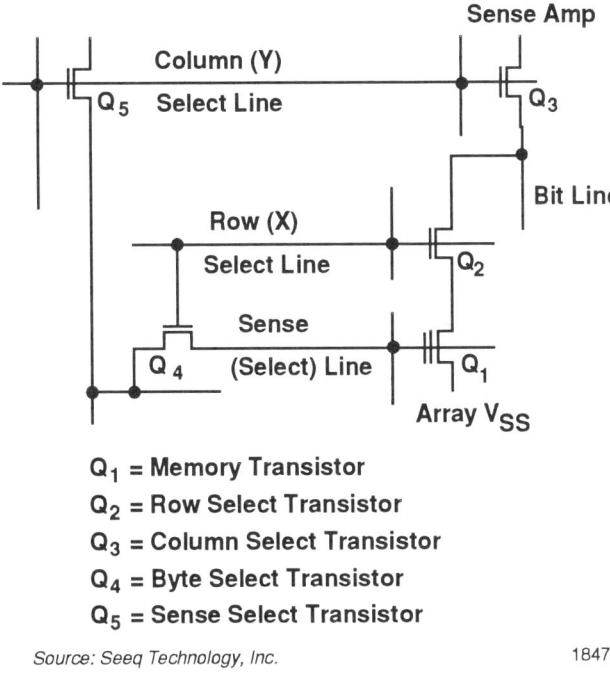

Q_1 = Memory Transistor
Q_2 = Row Select Transistor
Q_3 = Column Select Transistor
Q_4 = Byte Select Transistor
Q_5 = Sense Select Transistor

Source: Seeq Technology, Inc.   18475

Figure 1-13. Generic EEPROM Memory Cell

7. Flash

A modified version of the EEPROM called a Flash EPROM (or Flash EEPROM) is a compromise between the EPROM and the EEPROM. These flash devices are programmed like the EPROM and EEPROM, and can be erased electrically. The difference between the flash device and the EEPROM is that individual cells of the flash device cannot be erased. Rather, large groups of cells are erased as a block. This block size ranges from hundreds to thousands of cells. Figure 1-14 shows the cell organization of the flash device.

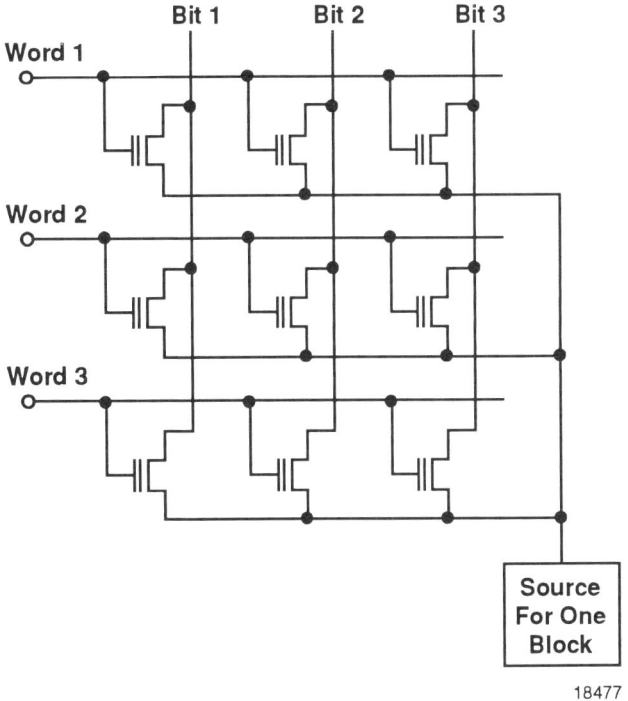

Figure 1-14. Flash EPROM Cell Connections

INTEGRATED CIRCUIT ENGINEERING CORPORATION

Products

The advantage of the flash device is that the potential cost per bit of storage approaches that of the EPROM, whereas that of the regular EEPROM will always be considerably higher due to the large cell size. The disadvantage of the flash device is that individual cells cannot be erased. Also, the flash structure suffers from the same limit to the number of erase cycles as the EEPROM. In the case of a the flash device, however, the limitation applies to an entire block rather than a single cell. This is due to the fact that the entire block must be erased when a single cell in the block is erased.

The early flash devices used a NOR structure. A structure that has been offered as a lower cost alternative is the NAND structure. The basic difference is in the layout of the individual cell. The NAND has an advantage in that it uses a smaller cell by stringing several EEPROM transistors together, without read/write contacts between each one. The disadvantage is that the peripheral circuitry required to read this type of transistor organization is complex, slow, and consumes a great deal of silicon. The slower (serial) readout NAND approach is sufficiently fast to be used as a replacement for rotating media, which is one of the major markets for flash devices.

The NOR connection used for many flash devices is also the standard connection used for all Mask ROMs, EPROMs, and EEPROMs. This configuration allows a fast read cycle, faster than most DRAMs. The terms NAND and NOR originate in the wiring of the cell transistors, which are connected together with common source and drain for the NOR, and in series, with the source of one cell transistor connected to the drain of the next, for the NAND. Figure 1-15 shows the cell arrangements.

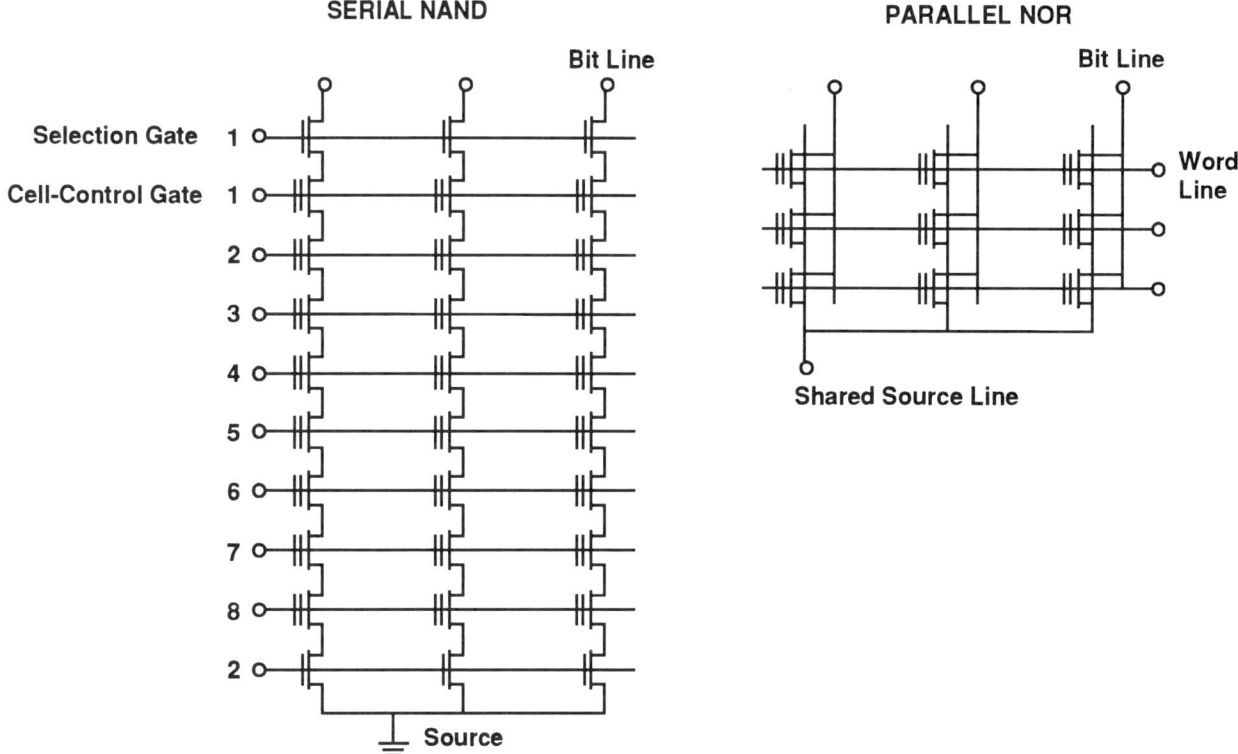

Figure 1-15. The NAND Versus NOR Contest

Any of these ROMs (MROMs, EPROMs, EEPROMs, or Flash) are logical places to store small computer programs. Because they are programs and not data, they are executed by the CPU, but never altered. A high-volume use of this is in microcontroller applications. Here the ROM can be either on the same chip as the CPU or a separate chip. Typical uses are appliance control and engine control in automobiles

The font cartridges in printers contain ROMs. In this case the ROMs contain the data needed to print particular fonts, giving the printer greater flexibility. Here, the data is stored but never changed.

8. Shadow and Battery-backed SRAMs (NOVRAMs)

Certain applications require the speed of normal memory, but have the additional requirement that the data must retained when the power is shut off, i.e., that they are non-volatile. Normal EEPROMs are non-volatile, but the write and erase times are far too slow for normal system operation.

One alternative that allows the use of SRAMs in these applications is combining SRAM and EEPROM on the same chip. In normal operation, the CPU will read and write data to the SRAM. This will take place at normal memory speeds. If the chip detects there is a power failure about to occur, the special circuits on the chip quickly (in a few milliseconds) copy the data from the SRAM section to the EEPROM section of the chip, thus preserving the data. The EEPROM "shadows" the data that is in the SRAM in normal operation. When power is restored, the data is copied from the EEPROM to the SRAM and operations can continue as if there were no interruption. Figure 1-16 shows the cell schematic of one of these devices.

Another alternative is the use of a small battery in the package of the SRAM chip. SRAMs can be designed to go to a sleep mode where the power consumed is very low but the data is retained. The combination of the SRAM and the small battery can be very cost effective, with retention times of five years. Note that notebook and laptop computers have this "sleep" feature, but here the regular system battery supplies the power.

B. STANDARD LOGIC

Standard logic was the first application of integrated circuits replacing discrete transistors. All early families used bipolar transistors and the names referred to the circuit design, such as Resistor-Transistor Logic (RTL), Diode-Transistor Logic (DTL), Emitter-Coupled Logic (ECL), and Transistor-Transistor Logic (TTL). These are shown in Figure 1-17.

Products

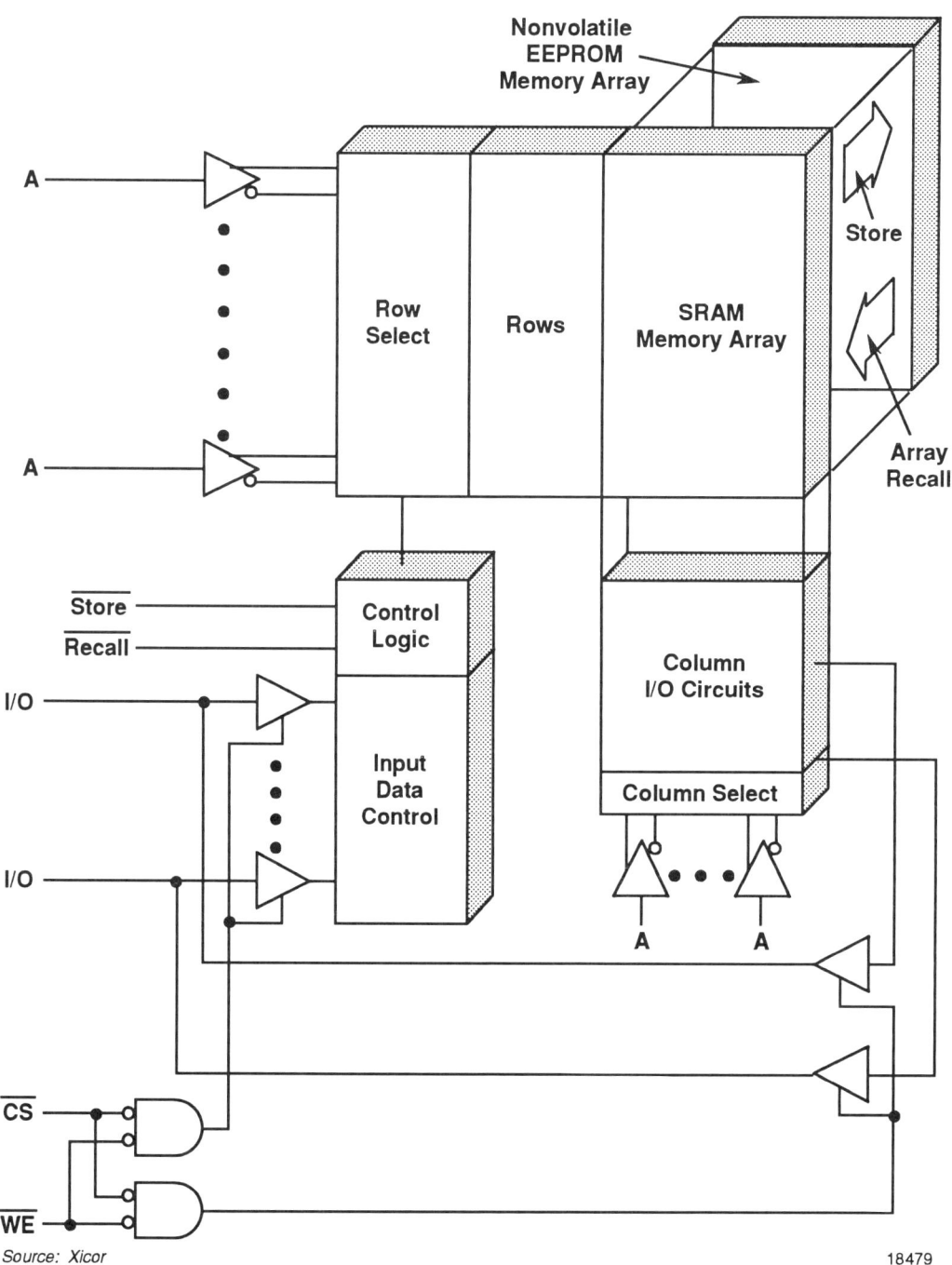

Source: Xicor

Figure 1-16. Block Diagram of the Xicor NOVRAM Family

Products

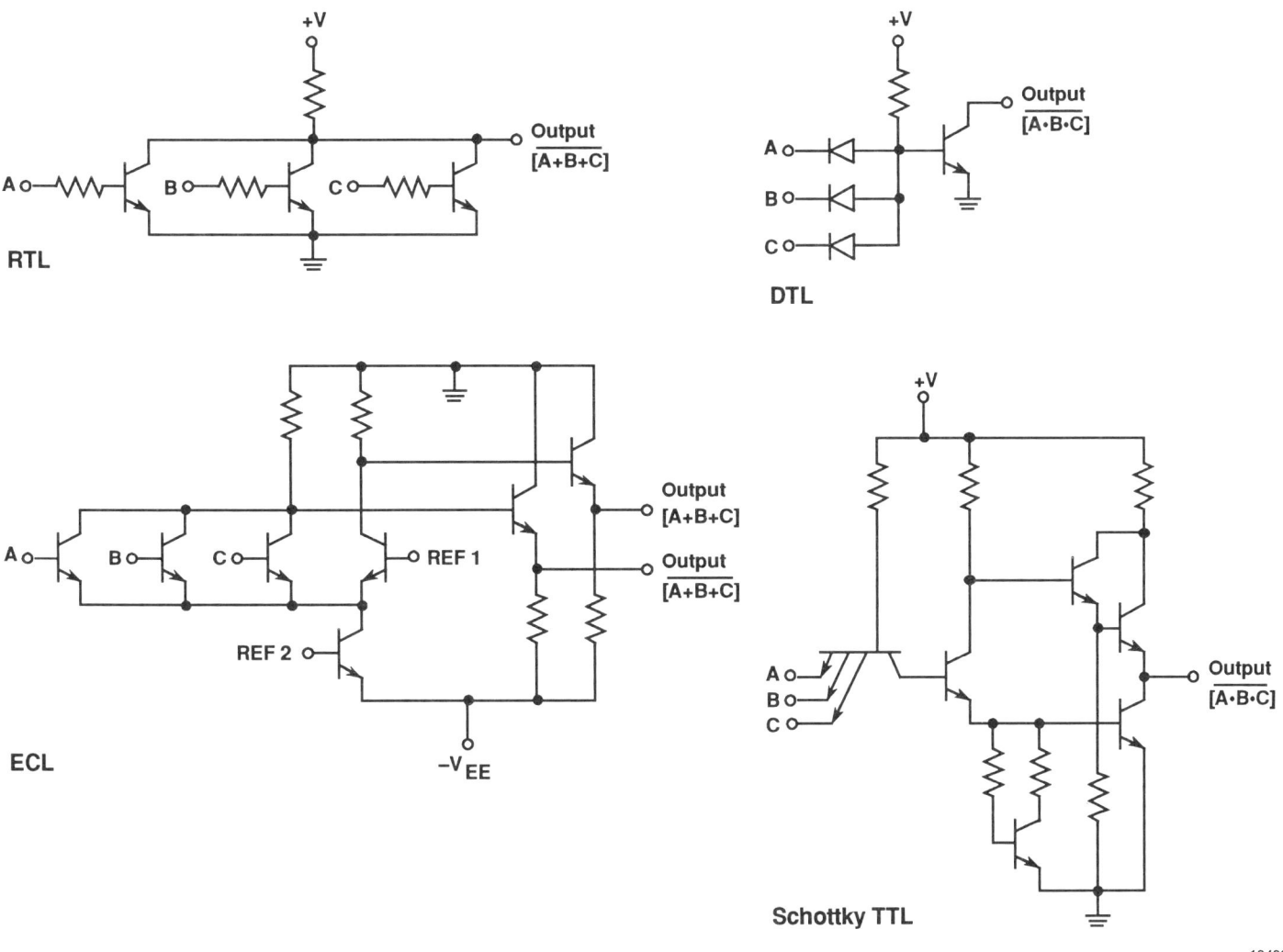

Figure 1-17. Various Standard Logic Circuit Designs

The most popular family was TTL, introduced by TRW and Sylvania in the early 1960's. The design was popularized by TI as the 54/74 series in about 1966. As the years passed, improvements in transistor performance allowed new families to be introduced that offered better speed or lower power, while retaining compatibility with the older system designs. Most manufacturers even used the same suffix numbers, as these were familiar to systems designers. Some of the families are manufactured using CMOS or BiCMOS processes, as well as the more familiar bipolar, offering improved performance or considerably greater functionality.

Figure 1-18 shows the various acronyms that name the families and the relative merits of each family in terms of speed and power.

Family Name (Common Reference Name)		Technology	Relative Speed	Relative Power Consumption/Gate
Schottky TTL	(Schottky)	Bipolar	4	7 (Highest)
Low Power STTL	(LS)	Bipolar	6	5
Advanced Schottky TTL	(AS)	Bipolar	1 (Fastest)	6
Advanced Low Power TTL	(ALS)	Bipolar	5	4
High Speed CMOS TTL	(HCT)	CMOS	7 (Slowest)	3
Advanced CMOS TTL	(ACT)	CMOS	3	1 (Lowest)
BiCMOS TTL	(BCT)	BiCMOS	2	2

Figure 1-18. Logic Families

Virtually all the parts in these families are general purpose logic, i.e., they are suitable for virtually any application that needs that particular logic function. Today, Very Large Scale Integration (VLSI) CMOS is designed to perform the function of hundreds of these chips, and for very specific applications.

C. MICROPROCESSORS AND MICROCONTROLLERS

1. Microcontrollers

The terms microcontroller, microcomputer, embedded controller, microprocessor, and CPU are used in the trade press without precise definition. ICE defines a microcontroller (MCU) as a device that contains all of the necessary functions of a computer on a single piece of silicon or other semiconductor material. These necessary functions include the central processing unit, Random Access Memory (RAM) used for read-write memory, nonvolatile program memory (ROM or ROM variations such as EPROM, EEPROM or flash), and at least one input/output port. The main thing that separates MCUs from microprocessors (MPUs) is that MPUs do not have on-chip ROM.

MCUs are sold almost exclusively as embedded controllers. Therefore, the statement "all MCUs are embedded controllers" is almost 100 percent correct. An embedded controller is a processor with its support circuitry whose function is to control the operation of some piece of equipment. However, all embedded controllers are not MCUs, as in many large systems regular MPUs are used as embedded controllers.

2. RISC vs. CISC Designs

The terms RISC (Reduced Instruction Set Computer) and CISC (Complex Instruction Set Computer) refer to the instruction vocabulary available to the programmer that will use the processor. The RISC concept was proposed by Patterson and Ditzel back in 1980 at IBM. The computer created was technically very different from the typical IBM computer of that timeframe.

The idea was to create a computer that was much simpler in design, and therefore could have other features that would allow the machine to operate considerably faster. As an example, the area on the processor chip taken up by the complex decoding required for the large repertoire of instructions in the CISC could be used for a large number of on-chip registers for the RISC, making the RISC very fast.

Today, the limitations on transistor count per IC chip no longer cause many significant design compromises. Many "good ideas" from the RISC world have been implemented in modern CISC designs. Likewise, some of the early RISC ideas have been found to be impractical and have been dropped.

Modern RISC chips will undoubtedly remain leaders in performance for two reasons. First, most modern CISC chips are designed to be compatible with the original designs first introduced in the 1970's, while the RISC chips are based on 1980's knowhow. Second, RISC chips can be applied to parallel applications much more easily, and this is a definite trend for all high-performance applications.

D. ASIC

The level of integration possible with Very Large Scale Integration (VLSI) caused a major change in the design of ICs. The capability of putting thousands of transistors on a single chip meant that more than one standard logic function could be placed on each chip. This, in turn, meant the system design would determine the circuitry of each IC chip. The IC could then be customized for a very specific application or product.

Application Specific Integrated Circuits (ASICs) are those ICs designed by or for the customer to fill a very specific function. Another type of IC designed for a specific application is the Application Specific Standard Product (ASSP). The typical differentiation between the two is that an ASSP is purchased by more than one customer, and it will typically have a datasheet describing the performance of the chip.

1. Custom

Several methods are used to convert a logic level design (a design composed of logic gates such as AND, OR, NAND, NOR, etc.) into the design of the IC chip. Full custom design refers to those chips designed at the transistor level, i.e., where each transistor is sized to fill the exact requirements. It is used for very high volume applications because the design cost is very high.

2. Standard Cell

Standard cell is the most common design approach used for VLSI ICs. The concept consists of two phases. First, "cells" or "macros" are designed to perform logic functions. These macros may be as simple as a two-transistor inverter or as complex as a several thousand transistor Arithmetic Logic Unit. The macros are characterized in terms of performance, power supply connections, input/output wiring requirements, etc. Figure 1-19 shows a "2:1 multiplexer" macro containing eight transistors. This would be considered a small macro.

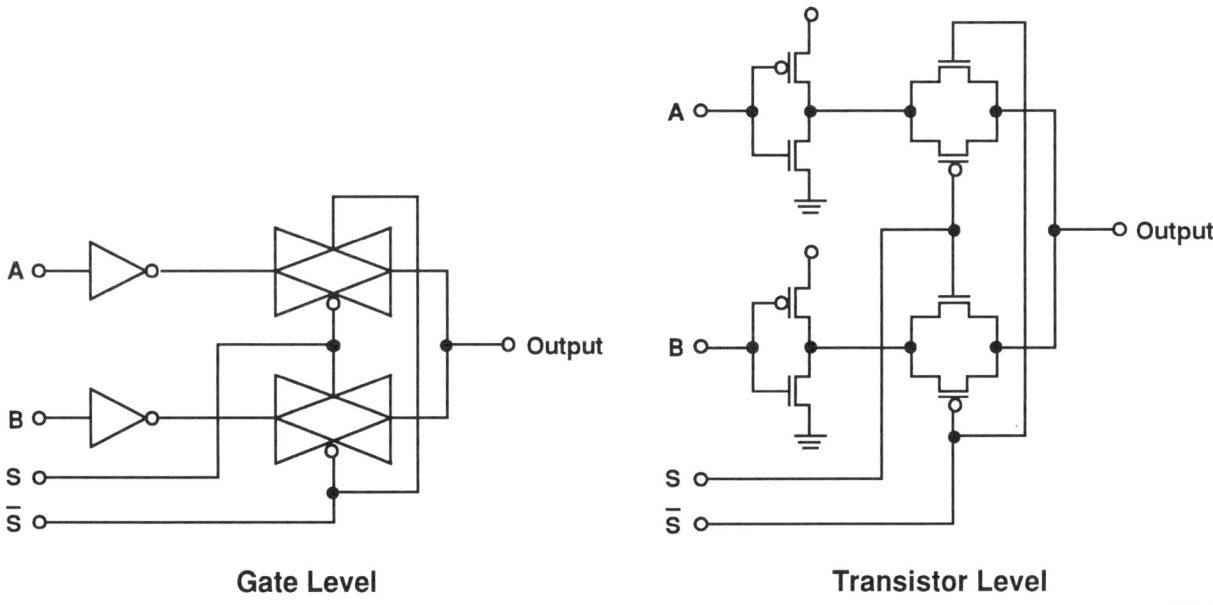

Figure 1-19. Two-Input Multiplexer

The second phase of a standard cell design is the actual design of the total chip. The designer uses the macros to create the total chip function. The computer then "places" the cells in an IC layout and connects them to achieve maximum performance. The standard cell approach is not as efficient as the full custom approach in terms of the amount of silicon used, i.e., the IC chip will be larger than with full custom. However, the standard cell approach requires considerably less engineering effort to design, as the macros can be re-used for different designs.

3. Gate Array

Another ASIC approach is the gate array. In this technique, only one basic cell is created, and is repeated on the chip thousands of times. The basic cell usually consists of four transistors as shown in Figure 1-20. These cells are not complete, requiring several more connections to become complete logic gates.

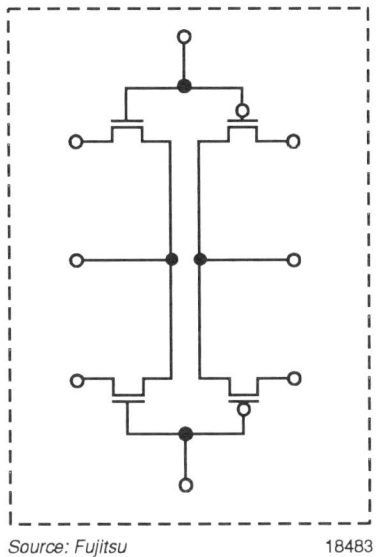

Figure 1-20. The Basic Gate Array Cell

The final metal layer(s) of the IC is used to complete these connections, using the transistors of the basic cell to create different kinds of logic gates. This is shown with the heavy connection lines in Figure 1-21. The gate array is less efficient in silicon usage than the standard cell, but has several advantages. One advantage is faster turnaround time in the IC manufacturing fab, as wafers can be partially processed and stored, awaiting only the interconnect metal layer(s). Another advantage is that new complex cells can be designed quite easily from the basic cell building blocks.

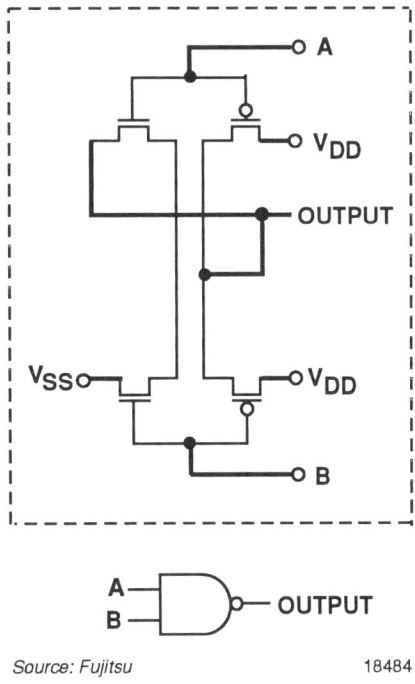

Figure 1-21. The Basic Gate Array Cell as a 2-Input NAND Gate

The chip designer uses almost the same design tools whether designing for a standard cell or gate array. The libraries for either methodology will be slightly different. The software that creates the actual chip is very different, but this is transparent to the designer.

Products

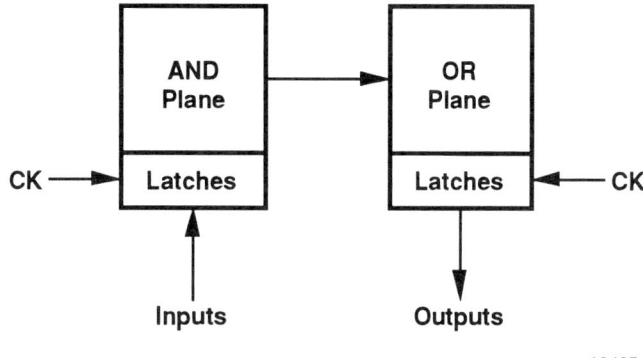

Figure 1-22. AND-OR PLA

4. PLDs

a. PLAs, PALs

The use of Programmable Logic Devices (PLDs) is another ASIC implementation approach. PLDs cover a broad range of IC chip categories. The earliest PLDs were implemented in a modified version of the general purpose Programmable Logic Array (PLA) shown in Figure 1-22, and named PAL by Monolithic Memories. These parts differed from the PLA in that the OR plane is fixed rather than infinitely variable. An example of the PAL architecture is shown in Figure 1-23.

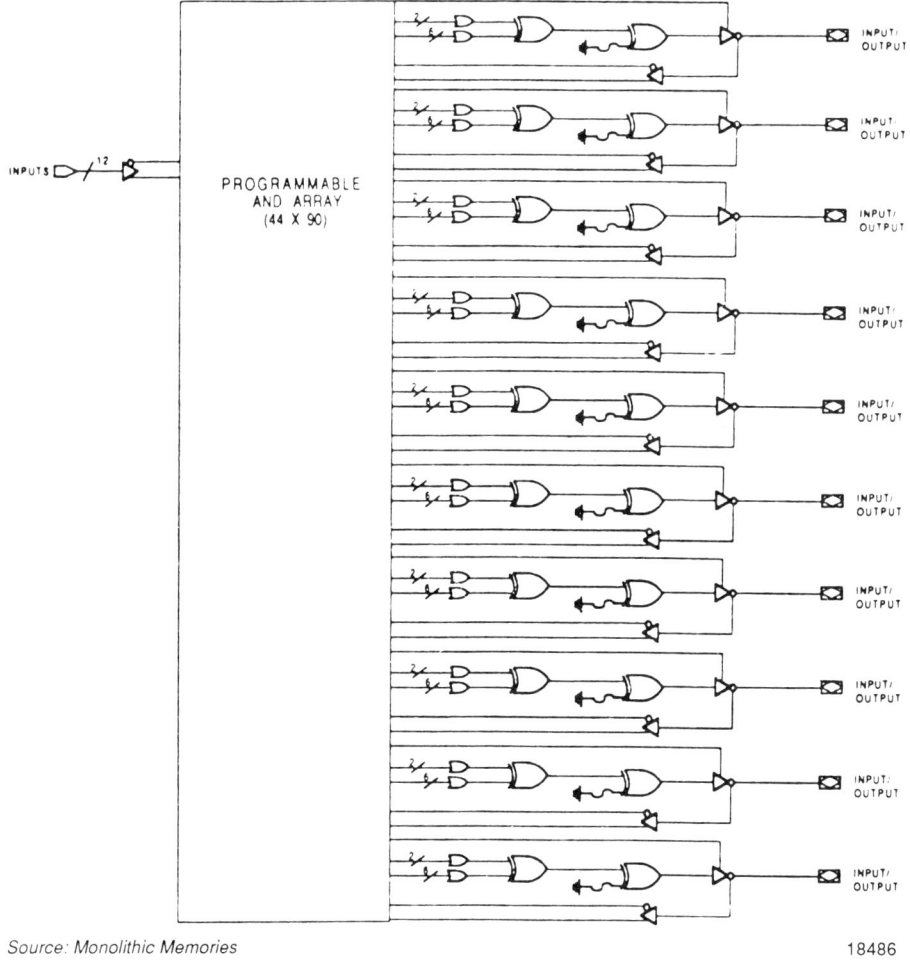

Source: Monolithic Memories

Figure 1-23. Block Diagram of an AmPAL22XP10

1-20   INTEGRATED CIRCUIT ENGINEERING CORPORATION

These early PALs were introduced in bipolar TTL logic, with metallic fuses that were "blown" to program the part in the desired logic configuration. Today, many of the new PAL devices are CMOS, using EEPROM transistors instead of the metallic fuses. In addition, the usefulness of the device has been enhanced by the addition of complex input and output "macros" (not to be confused with the macros or cells in the standard cell methodology). These macros are programmed to form a variety of logic functions. Figure 1-24 shows a common output macro for a widely used PAL device.

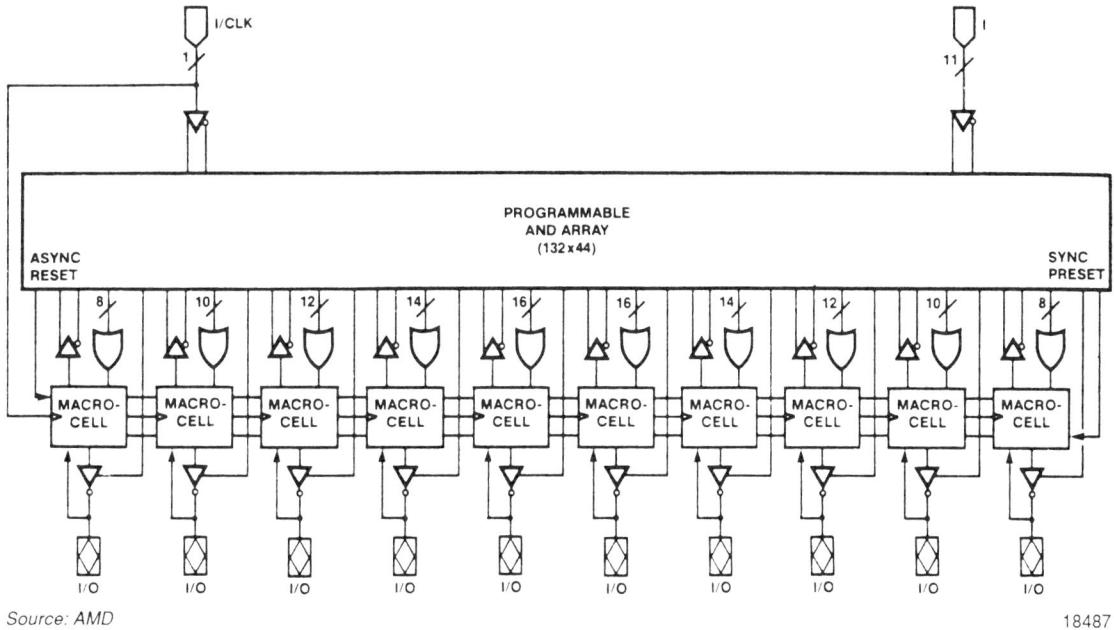

Figure 1-24. PAL Block Diagram

b. PGAs

In the late 1980's, new devices called Programmable Gate Arrays (PGAs) were introduced. These structures use proprietary internal architectures, and can be thought of as PALs with very small input AND and OR arrays, but with a very large quantity of output and input macros. An example of this is shown in Figure 1-25.

Although the parts don't actually resemble an array of gates as the name implies, the user designs the circuit as if it is. The "macros" are programmed by the vendor's software to implement the user's requirements. The complex part of the system is the software, as it must optimize the use of the macros very carefully to fully utilize the IC chip. The metallization on the chip also tends to be very complex. It must not only allow the macros to be connected to one another with as many paths as possible, but must also supply paths for the programming logic.

Products

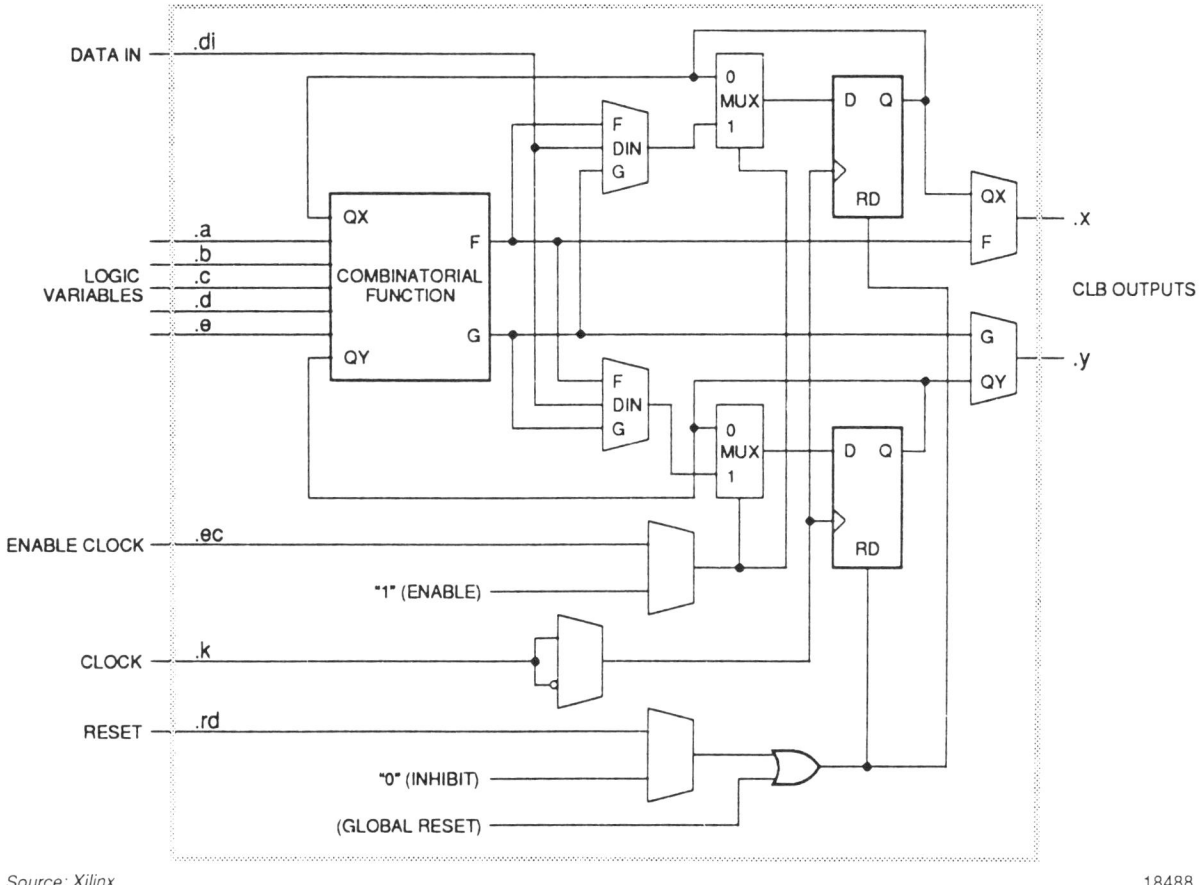

Source: Xilinx

Figure 1-25. Configurable Logic Block

The storage of the programming for each macro can be stored in a number of different cells. Flash EEPROM, EPROM, SRAM, and anti-fuse cells are all used for storage of the program configuration. The SRAM is the only storage method requiring the program to be temporarily stored in another chip or on disk, as the SRAM cells are volatile and lose their configuration when the power is removed.

Anti-fuses are special layers or films that change electrical properties when pulsed with some electrical stimuli, such as a voltage pulse. In most cases the original state of the film is a non-conductor. After the electrical stimulation, the film becomes an electrical conductor. Since this is the opposite effect to that of a metallic fuse, the term anti-fuse is used to describe the structure. In nearly all cases, the structure acts as a resistor.

E. ANALOG

Analog (or linear — the names tend to be used interchangeably) is a general category of ICs that includes all ICs that do not fit into one of the digital IC categories. In most cases, the ICs are designed to operate at higher voltages than the standard digital familes, and in some cases they do perform analog functions, as described below.

1. Amplifiers

Amplifiers are circuits that are a combination of active devices (such as transistors and diodes) and passive devices (capacitors, resistors and inductors) that are designed to increase electrical signals from a low level to a high level. The circuits in a radio or TV that increase the signal level from that transmitted over the airwaves to a signal detectable by the human ear are audio amplifiers, i.e., they amplify audible electrical signals. Figure 1-26 shows an amplifier with input and output signals, both with low and high amplification settings.

Many industrial applications use general purpose amplifier chips that are called operational amplifiers, a term used to denote a general purpose amplifier with infinite gain (the amount of amplification available) and very high input resistance (causing little or no effect on the signal connected to the amplifier input). Figure 1-27 shows a small amplifier schematic.

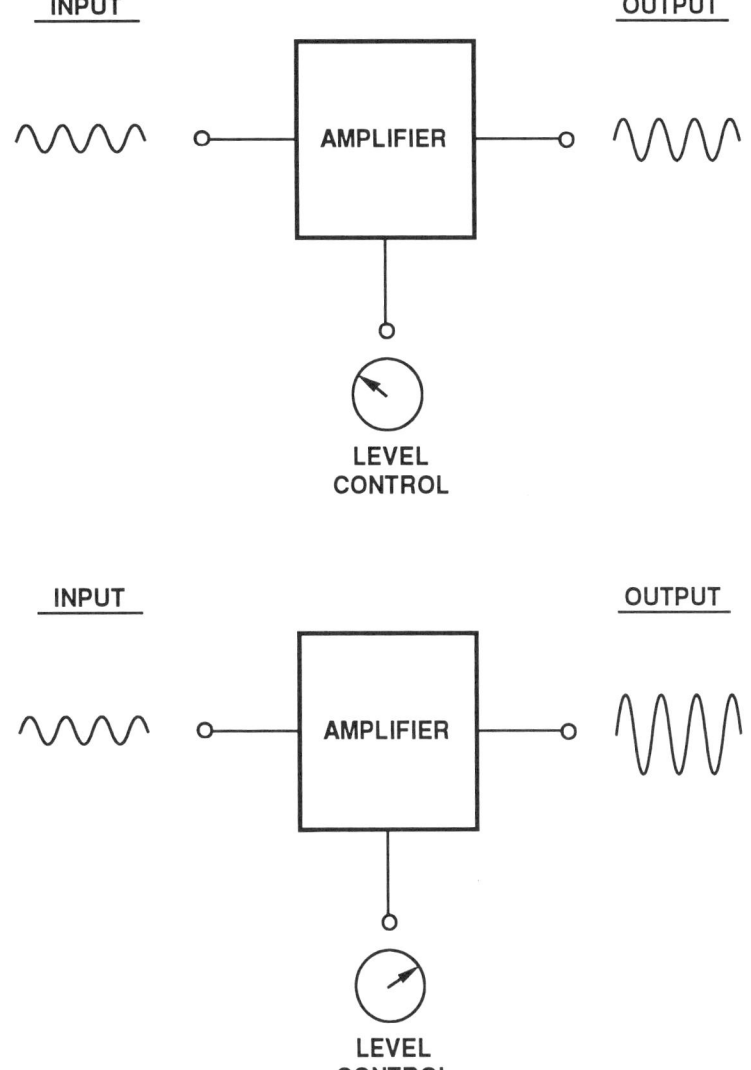

Figure 1-26. Analog Example

Although a simple amplifier could be built with three of four transistors, the design would not be very practical. Semiconductor devices change dramatically with changes in temperature, and these changes must be nullified with compensation circuits. This accounts for the large number of devices in the figure.

2. Voltage Regulators

Voltage regulators are circuits designed to deliver very precise electrical voltages to other ICs in a system. The input voltage to a voltage regulator may vary over 25 percent, while the output is controlled to less than one percent. There are thousands of different designs used for the various types of equipment in use. The key to high-quality voltage regulators is the ability to tolerate large fluctuations on the input, while maintaining tight control on the output, and to do so over a wide temperature range.

Products

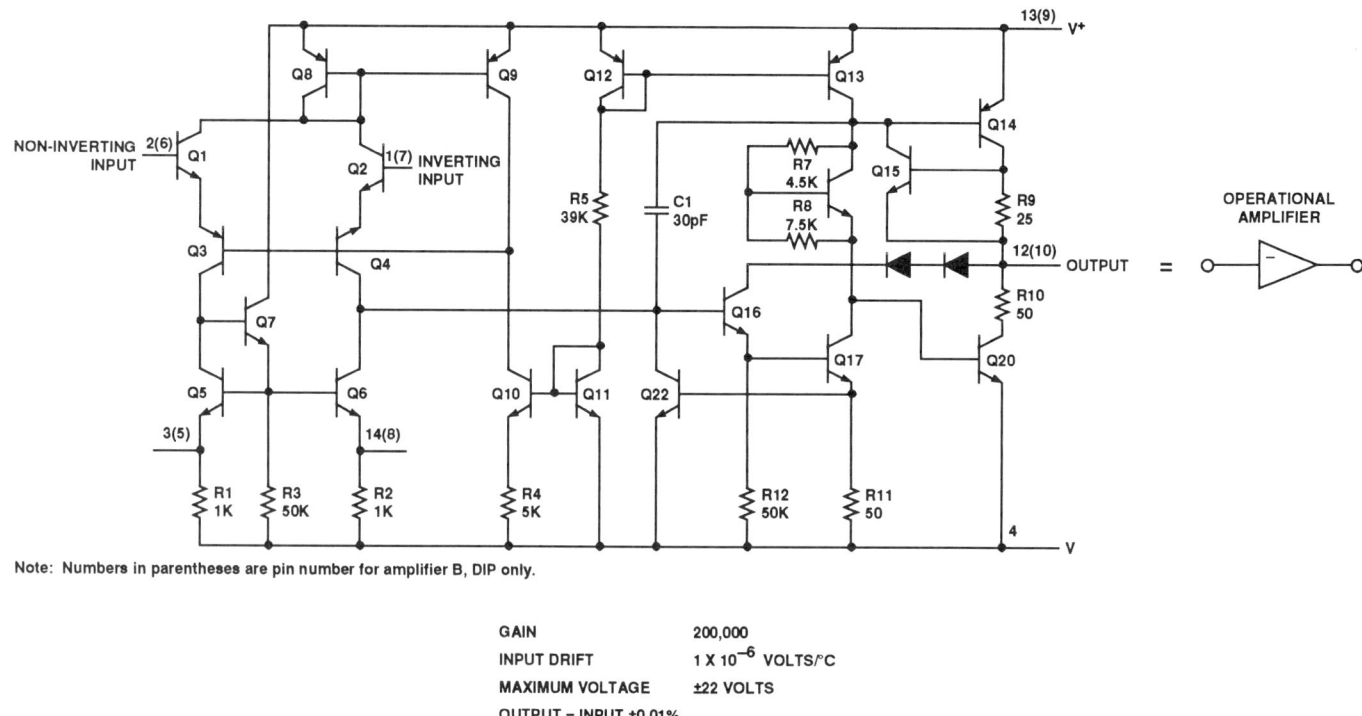

Note: Numbers in parentheses are pin number for amplifier B, DIP only.

GAIN	200,000
INPUT DRIFT	1×10^{-6} VOLTS/°C
MAXIMUM VOLTAGE	±22 VOLTS
OUTPUT = INPUT ±0.01%	

Figure 1-27. Precision – Linear

3. Data Conversion

Data conversion circuits are those that translate from analog signals to digital signals, and back again. An example of this is the typical telephone call between cities. The voice (an analog signal) is converted to a digital signal at the local telephone office. The digital signals are sent from the local office to the office near the person called. The digital signals are then converted back to analog signals for the person called to hear.

Figure 1-28 shows a typical voice or music waveshape. This signal is converted to digital data by sampling the waveshape at specific intervals. These sample points and values are shown as bars and numbers on the figure. This "digitization" function is performed by an analog-to-digital (A/D) converter.

The digital values are transmitted, in order, along the transmission media, wire, optical fiber, or radio link, to the receiving station. At the receiving station the digital data is sent through a digital-to-analog (D/A or DAC) converter to recreate the analog signal. Figure 1-29 shows the relative functions of these devices.

Products

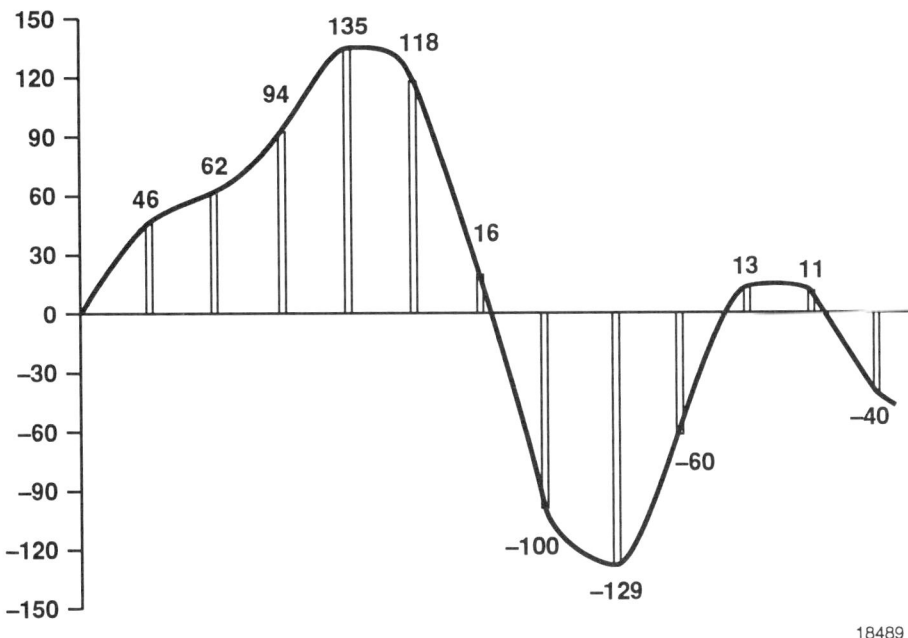

Figure 1-28. Sampled Analog Waveshape

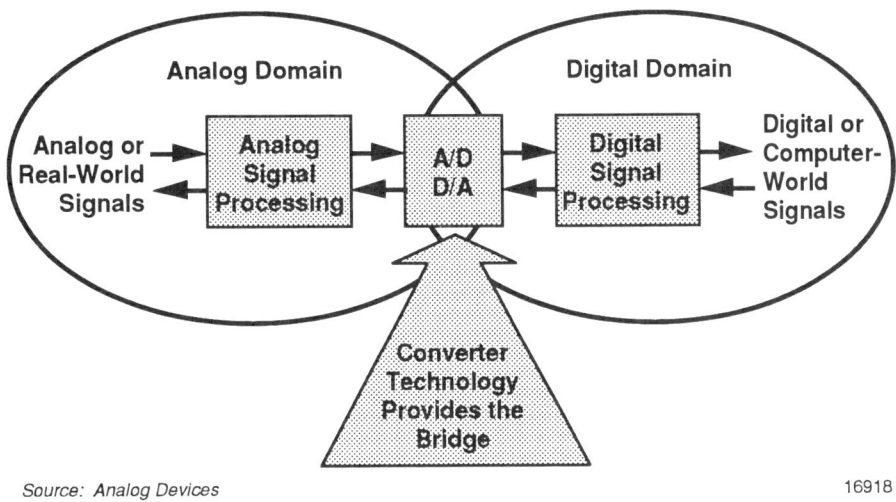

Figure 1-29. Real-World Signal Processing

A special version of these converters is the CODEC found in telephone systems. These ICs contain both A/D and D/A converters, so they support bidirectional communications. The devices follow very special rules in conversion so various brands of equipment are able to communicate.

INTEGRATED CIRCUIT ENGINEERING CORPORATION

Products

The advantage of the digital transmission method for long distance (versus the 1950's vintage "all analog" transmission) is the quality of the signal at the receiving end. Analog signals degrade (become lower) over distance, but the noise levels tend to stay constant. Therefore at the receiving end the signal becomes mostly noise. When digital data is transmitted over long distances, the power of the signal degrades, but the "ones" remain distinguishable from the "zeros," so a noise-free version of the starting signal can be reconstructed at the receiving end.

4. DSP

Digital Signal Processor (DSP) ICs are special microprocessors designed for very high speed mathematical computations. The applications are those where the inputs and outputs are both analog, but where the processing can be digital or analog. Historically, the processing was performed by analog components because of cost. However, as the cost of digital processing has dropped, many applications began using DSP chips because of the improved performance.

In these applications, the "real world" analog signals are converted to digital data by an A/D converter. The digital information is acted on by the DSP chip and converted back to analog by a D/A converter. Figure 1-30 shows the DSP chip with the two converters and the program and data memory.

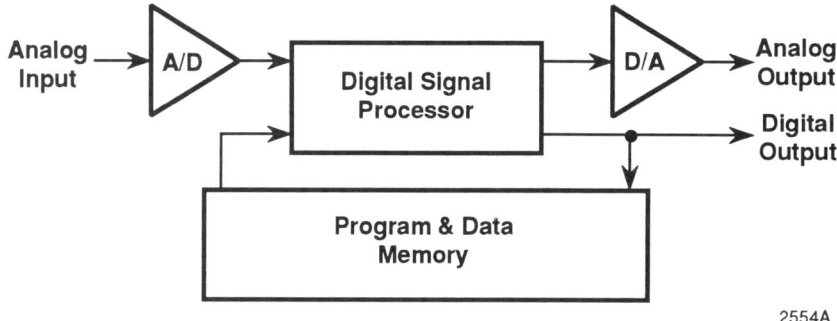

Figure 1-30. Digital Signal Processing

5. Interface Circuits

Interface circuits are a very general category of ICs that are fully digital, but act as translators between two or more signal levels. For example, most computers switch between 0 volts and +5 volts internally, logic zero and logic one, respectively. To communicate with serial printers using RS-232C interface, the signals must be changed to a signal varying between +12 volts and -12 volts. A single IC accomplishes this translation.

Subscriber Line Interface Circuits (SLICs) are used in telephone switches (at the phone company or local PBX) to connect to the telephone lines (actual wires) from the telephone sets. These devices perform the interface functions such as recognizing when the telephone user has taken the phone off-hook and wants a dial tone, passing the dial tone to the user until the first digit is dialed, passing the ringing or busy signal to the user, etc.

Products

These ICs must be able to withstand a significant amount of "abuse" in the form of electrical transient signals, poor wiring, etc., where signal levels are in the 50-volt range, and also communicate on a logic level with the electronic circuits in the telephone switch, usually between 0 volts and +5 volts.

6. Analog Arrays

Analog arrays are an analog version of digital gate arrays, or as close as the idea can get. The ICs are groups of discrete transistors, resistors and capacitors that can be interconnected with metallization to form some type of analog or mixed-mode circuit. These circuits require the customer (or whoever designs the final circuit) to understand linear design. The advantage over standard analog is that special custom requirements can be designed into the circuit, without requiring a full custom design. The disadvantage is that the array approach uses fixed component sizes, therefore limiting the range of performance compared to a full custom approach. Figure 1-31 shows an analog array chip. The list in Figure 1-32 shows the various components on the chip available to the designer.

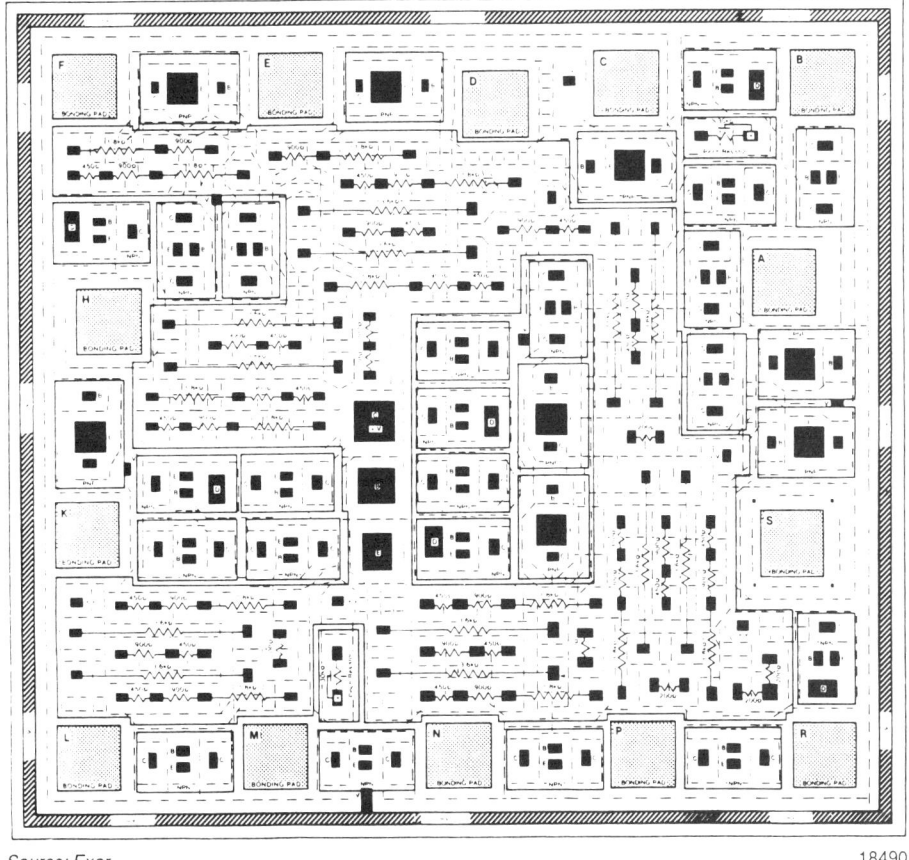

Source: Exar 18490

Figure 1-31. Analog Array

Total Component	110
Bonding Pads	14
Operating Voltage	20V
NPN Transistors	23
PNP Transistors	8
Schottky Diodes	6
Resistors:	
200Ω	8
450Ω	18
1,800Ω	20
3,600Ω	12
30,000Ω	2

Figure 1-32. Analog Array Component List

7. Mixed-Mode

Mixed-mode ICs are special cases of full custom or standard cell ASIC products that include both digital and linear circuitry on the same chip. These chips are becoming popular as VLSI technology moves towards single-chip solutions.

The process technology for mixed-mode ICs requires certain compromises. The performance of a single mixed-mode chip is usually less than the equivalent function performed by an analog chip and a digital chip working together. However, the cost of a single chip is usually much less than the two-chip approach.

F. SYSTEM TERMS

1. LANs and Ethernet

Local Area Networks (LANs) are used to connect computers together. The typical wiring for a LAN differs from the old method of connecting each terminal or small computer to a host (usually a mainframe or minicomputer). The LAN allows any individual device connected to the LAN to communicate with any other device, without going through a central computer. A good example of the power of a LAN is the sharing of expensive high-speed printers.

Ethernet is a specific protocol, or set of rules, for both the software and hardware that is used to create a LAN. These rules were established in the mid-1970's by Xerox, DEC, and Intel. The rules formed a standard, enabling different companies to supply equipment that would be able to send data back and forth.

2. Displays - Active Matrix LCDs

Liquid Crystal Devices (LCDs) are visual display devices that operate by blocking the transmission of polarized light. The liquid crystal material itself is a very thin liquid film sandwiched between two glass plates. When an electric field is applied to the liquid, the liquid changes from an orderly pattern that twists the polarized light to a different orderly pattern that causes the light to be absorbed.

These displays are used in simple applications such as calculators, watches, instrument displays, etc. In these applications, each segment (usually seven segments are used for the ten numbers) is connected directly to the IC chip that applies the full battery voltage or zero volts to activate the display.

Other applications, such as replacements for Cathode Ray Tubes (CRTs) in TV sets or computer displays, are more difficult for LCDs. In these cases the voltage used to drive the LCD on or off is shared with a portion of the cells in the same rows and columns as the cell being addressed.

To make this address mechanism more effective, a transistor is needed for each pixel of the display, and it must be located on the glass of the display. These displays are referred to as Active Matrix LCDs. Silicon is deposited on the glass and thin-film transistors are manufactured in this material. The contrast of these displays is considerably better than regular LCDs, but the manufacturing cost is considerably higher.

Products

2 BASIC INTEGRATED CIRCUIT MANUFACTURING

A. STARTING MATERIAL

Silicon is one of nature's most useful elements. Silicon is the material most commonly used for the manufacturing of semiconductors. Silicon, as a pure chemical element, is not found free in nature. It exists primarily in compound form with other chemical elements. In all of its various forms, silicon makes up 25.7% of the earth's crust, and is the second most abundant element in the Periodic Table Of Elements. It is exceeded only by oxygen. Silicon occurs chiefly as a compound of silicon and oxygen called an oxide or as a compound of silicon and salts called a silicate.

Silicon in the form of an oxide most commonly occurs as silicon dioxide, SiO_2, generally called silica, or sand. Other common forms of silicon dioxide are quartzite, quartz, rock crystal, amethyst, agate, flint, jasper, and opal. Many of the previous forms contain minute quantities of other elements that give the forms color. Some of these minerals are known as semi-precious gem stones.

Granite, hornblende, asbestos, feldspar, clay, mica, etc., are but a few of the numerous silicate minerals. Silicon, as sand, is one of the main ingredients of glass. Silicon is an important component in steel, aluminum alloys, and other metallugurical products. Silicon carbide is one of the more useful abrasive materials.

1. Purification

Silicon for semiconductor applications is taken from quartzite, the rock form of silicon dioxide. Quartzite has trace levels of other elements that must be removed in the purification process. The purifying of quartzite consists of reacting the quartzite with some form of carbon material. The carbon may be in the form of coal, coke, or wood. The process is performed at a temperature of approximately 2000°C. This chemical reaction produces metallurgical grade silicon (MGS) that is 98% pure, which is not good enough for semiconductor use, so must be further purified. This silicon is reacted with very strong hydrochloric acid (similar to strong swimming pool acid) to form a new liquid called trichlorosilane. The liquid is then purified by fractional distillation (similar to the distillation process to make whiskey). The resulting, highly purified trichlorosilane liquid is converted to polycrystalline electronic grade silicon (EGS) by the Siemens' process. The Siemens' process changes the liquid into a solid polycrystalline silicon (usually called polysilicon) rod. These process steps are summarized in Figure 2-1.

Basic Integrated Circuit Manufacturing

$$SiO_2 + 2C \xrightarrow{\text{Heat} \atop 2{,}000°C} Si + 2CO$$
Quartzite ; Coal, Coke, or Wood Chips ; MGS 98% ; Carbon Monoxide

$$Si + 3HCl \xrightarrow{\text{Heat} \atop 1{,}200 - 1{,}300°C} SiHCl_3 + H_2$$
MGS Powder ; Gas ; Trichlorosilane

Remove impurities by fractional distillation

$$2SiHCl_3 + 2H_2 \xrightarrow{\text{Heat} \atop 1{,}100°C} 2Si + 6HCl$$
Gas ; Gas ; Solid EGS Poly ; Gas

Silicon - 25.7% of the earth's crust

Figure 2-1. Polysilicon Creation

2. Czochralski Crystal Growing

The next process step converts the very pure silicon from a polysilicon crystal form into a single crystal or monocrystalline form (Figure 2-2). This process is known as Czochralski crystal growing, often called Cz, the abbreviation for Czochralski.

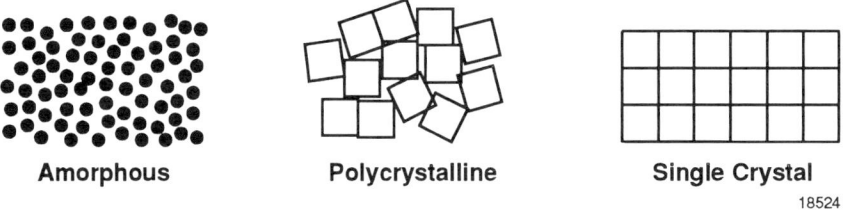

Figure 2-2. Types of Silicon Structures

The Cz process involves a special crystal growing furnace (Figure 2-3) wherein the pure polysilicon is placed into a very pure quartz crucible, along with the dopant material to make the wafer N-type or P-type. The process chamber is evacuated and purged with very pure argon. The crucible is heated to the melting point of silicon, 1420°C. A previously loaded monocrystalline silicon "seed" of the desired crystal orientation is lowered into the molten silicon and then slowly withdrawn as the "seed" and crucible rotate in opposite directions. This process causes the molten silicon to freeze out onto the "seed" crystal, forming a monocrystalline silicon ingot. The dopant species added to the polysilicon material determines the crystal's electrical characteristics.

The growing process is very automated and yields quality crystals. The growth process is very slow, typically 0.5 inch per hour for 150mm diameter crystals. Because of slow growth rates, the manufacturing process consumes large quantities of electricity.

Basic Integrated Circuit Manufacturing

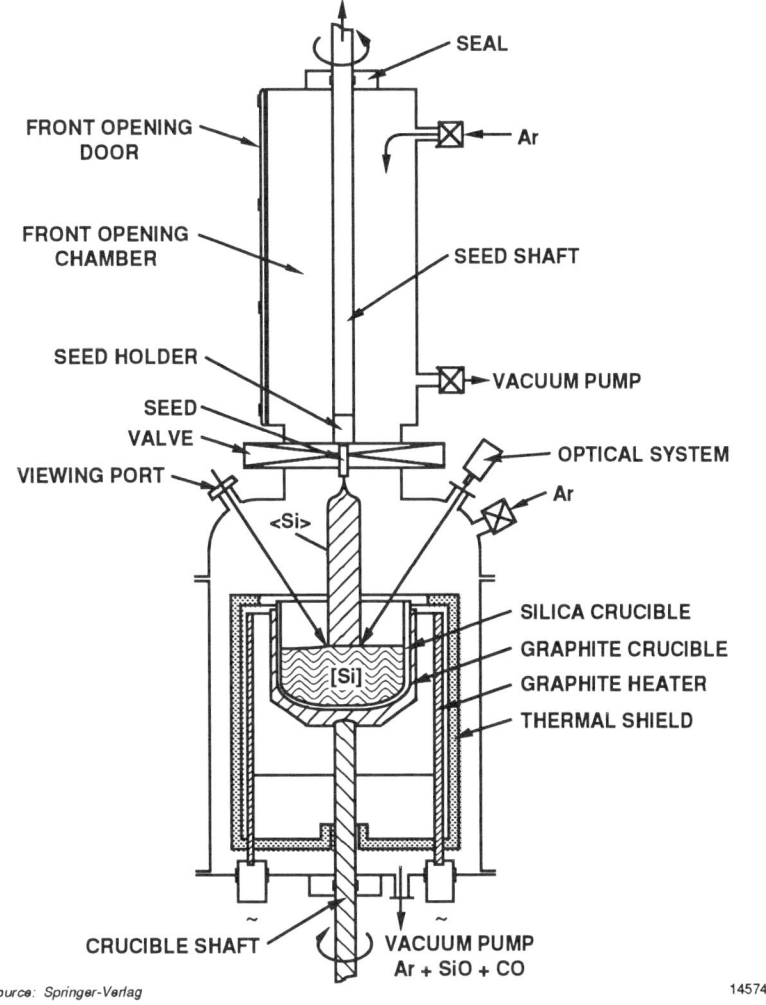

Figure 2-3. Czochralski Silicon Crystal Grower

After the growing process is completed, the silicon ingot is evaluated for both electrical and mechanical parameters. Generally, the yield is high and the usable silicon efficiency is 80 to 90 percent of the original crucible charge. This helps keep the cost down.

The diameter of the ingot during crystal growing cannot be controlled to the tolerance required by wafer handling equipment. To meet the diameter tolerance, the ingot is centerless ground to the exact diameter. Another part of the grinding process is to grind a major flat and minor flat or flats for crystal plane designation. The major flat is used for wafer orientation in wafer handling equipment and as a visual indicator to the manufacturing personnel.

3. Sawing Crystal Into Wafers

The next sequence of process steps involves sawing the ingot into the individual wafers and edge grinding the outer edge of the wafer circumference to a controlled shape. The sawing concept is illustrated in Figure 2-4. The edge grinding process removes the sharp edges of the wafer and dramatically reduces silicon particles when the wafer is handled.

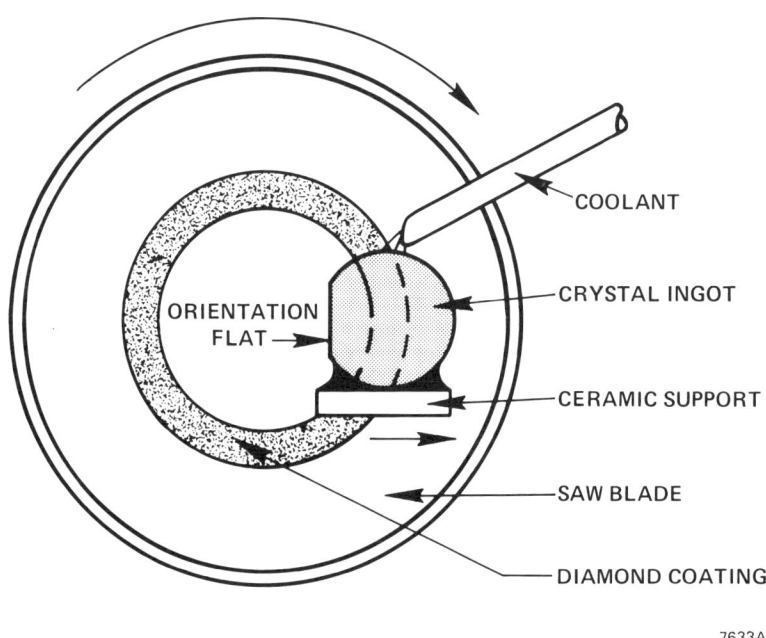

Figure 2-4. Inner-Diameter (ID) Sawing of Silicon Ingot

4. Wafer Preparation

The wafer is double-side (both sides simultaneously) lapped to reduce the surface roughness and then chemically etched to further smooth both sides of the wafer surface. A final surface polish is done on the designated side of the wafer. The designated side is determined by the ground wafer flat(s). The final surface must be very smooth (like a mirror) and free of surface defects and imperfections. A final chemical cleaning process will remove the polishing materials, particulants and any other potentially contaminating materials. These process steps are illustrated in Figure 2-5. Electrical and mechanical evaluation completes the processing.

The wafers are packaged in an ultra-clean environment and sealed in the storage-shipping containers. They are ready for use in the fabrication process.

Basic Integrated Circuit Manufacturing

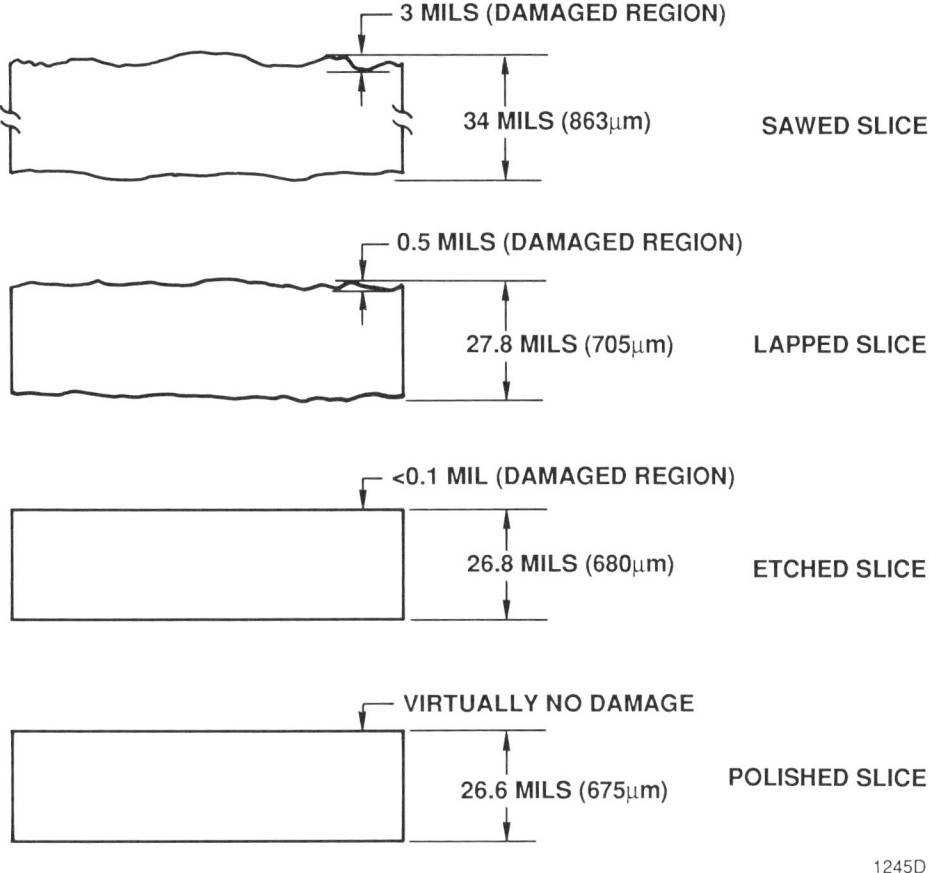

Figure 2-5. Wafer Preparation (for 150mm)

B. WAFER CLEANING

Contamination control during IC manufacturing is a major factor for Yield, Cost, Reliability, and Quality. Therefore, the total manufacturing cycle *MUST BE* controlled continuously to meet these needs. This means there is control over the environment, materials, chemicals, people, equipment and the process interactions with all of the above.

The physical size of contamination can vary over a large range. Figure 2-6 illustrates some of the contaminating materials and sizes. Contamination in any form, films or particles, larger than 10 percent of the smallest line width is considered detrimental. This is illustrated in Figure 2-7. In general, anything on the wafer surface that is not designed to be there is considered contamination.

An extremely critical part of the manufacturing sequence is the cleaning of the wafer surface after certain process steps and prior to other process steps.

Basic Integrated Circuit Manufacturing

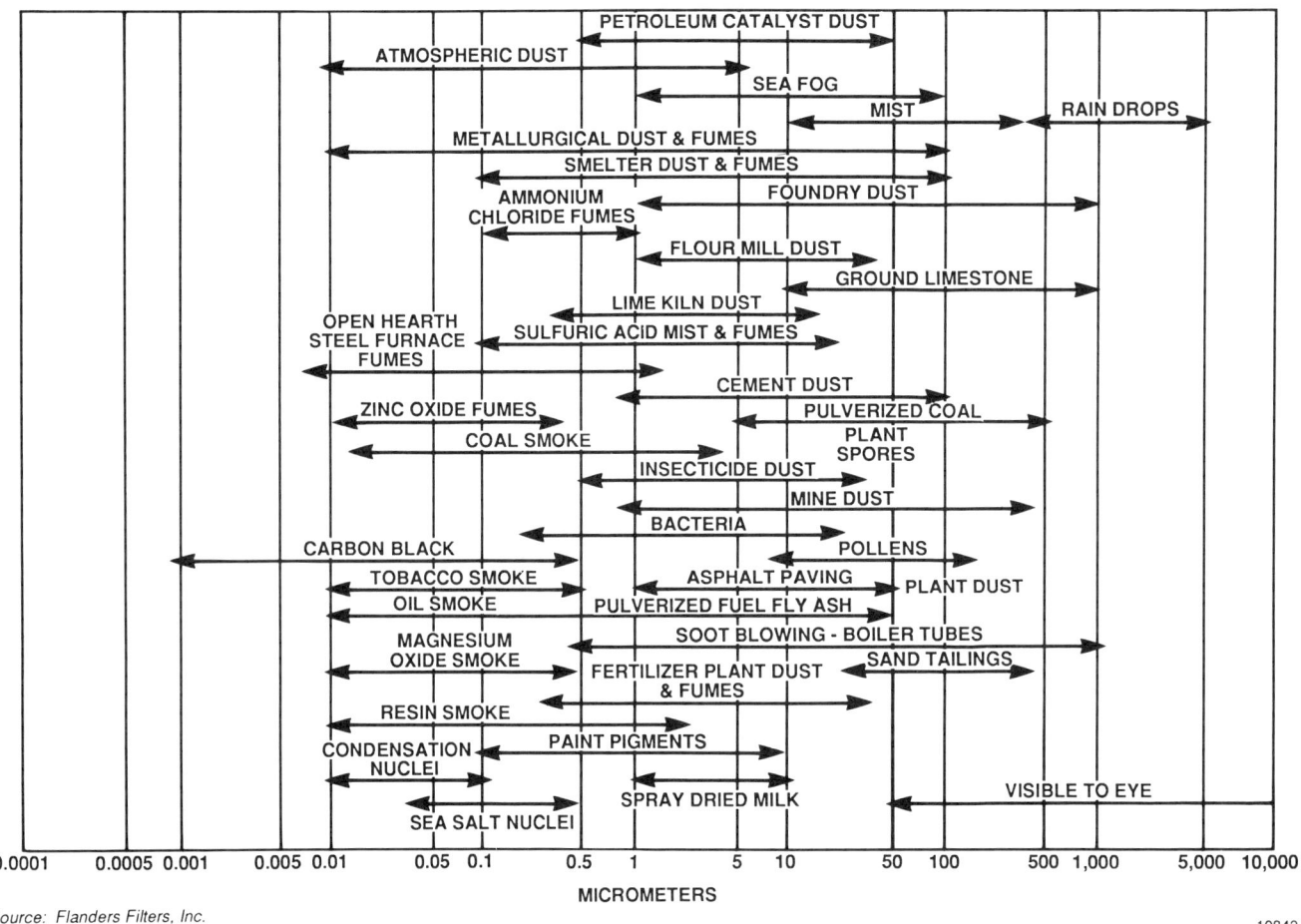

Figure 2-6. Sizes of Airborne Contaminants

Cleaning processes are required before the wafers are introduced into any elevated temperature process. A cleaning process is necessary to remove any form of surface contamination that may create a defective transistor within an IC die or produce instability in the circuit during the lifetime of the IC. The purity of the wafer surface is essential.

Contaminating materials on the wafer surface can lead to some of the following problems.

1. Prevent or mask effective cleaning or rinsing.
2. Prevent or mask effective concentration of dopants to be introduced into the silicon, whether by diffusion or ion implantation.
3. Cause poor or no adhesion of deposited layers.
4. Cause undesired chemical reactions and lead to decomposition of materials.
5. Alter the silicon crystal structure causing undesired electrical parameter changes.
6. Lead to long-term instability of electrical parameters.
7. Cause film degradation or catastrophic device failure.

Basic Integrated Circuit Manufacturing

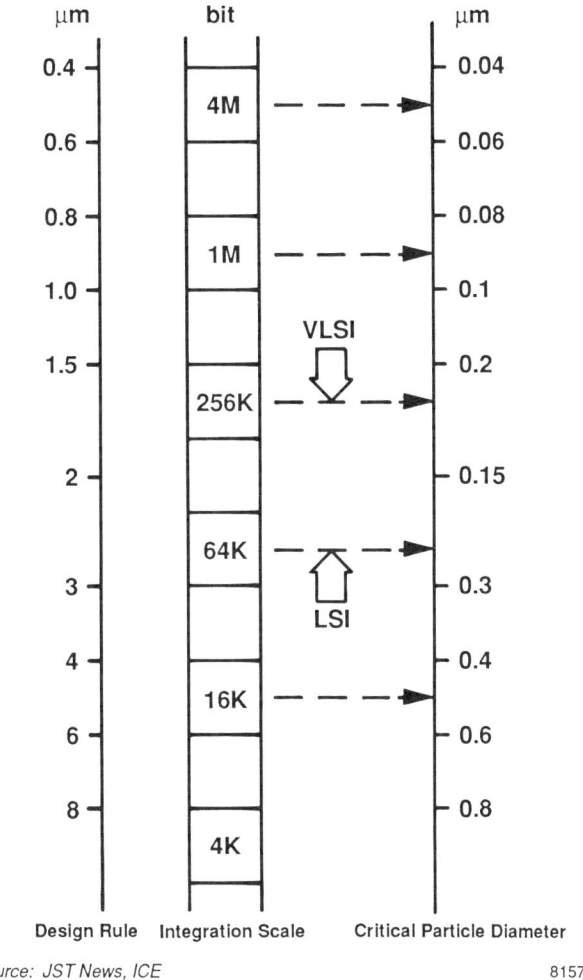

Figure 2-7. Correspondence Between Scale of Integration (Design Rule) and
Size of Particles to be Removed.

Many chemical cleaning processes have been evaluated for cleaning silicon wafer surfaces. These processes must be divided into two categories: those processes used prior to the metallization process and those used after the metallization process because of chemical attack on metal.

The selected process must remove a variety of contaminates and leave nothing on the surface as the result of the chemical reaction. This requirement has lead to the use of hydrogen peroxide (H_2O_2) as the preferred cleaning solution. The peroxide is a 30 percent concentration and unstabilized chemically. The unstabilized form is necessary because of purity.

With peroxide as the starting solution, two different chemical cleaning processes have evolved. One is known as the RCA clean. The other cleaning process is known by several different names: Piranha, Caro, and Sulfuric/Peroxide. These cleaning processes are summarized in Figure 2-8.

RCA TYPE CLEANING	SULFURIC ACID/PEROXIDE CLEANING (PIRANHA) (see note below)	
6:1:1 of H_2O; 30 W/W% H_2O_2; and 29 W/W% NH_4OH (as NH_3) followed by a mixture of 6:1:1 of H_2O; 30 W/W% H_2O_2; and 37% HCl. Keep both mixtures at 75° - 80°C for 15 minutes. UPDI rinse to 15 meg·ohm; spin/rinse/dry.	**CLEAN A**	
	H_2SO_4/H_2O_2 clean UPDI/quick dump rinse	10 minutes to specified resistivity
	50:1 etch (H_2O:HF) UPDI/quick dump rinse	60 seconds to specified resistivity
	UPDI rinse to 15 meg·ohm Spin/rinse/dry	
	CLEAN B	
	H_2SO_4/Ammonium persulfate UPDI/quick dump rinse	10 minutes to specified resistivity
	50:1 etch (H_2O:HF) UPDI/quick dump rinse	60 seconds to specified resistivity
	UPDI rinse to 15 meg·ohm Spin/rinse/dry	

Note: Stabilize H_2SO_4 acid at 80°C before adding H_2O_2.
Add H_2O_2 to make a 4:1 (H_2SO_4:H_2O_2) solution.
Solution temperature will rise to approximately 180°C after adding H_2O_2.

Figure 2-8. Wafer Cleaning Techniques Prior to Thermal Processing

The RCA clean is a two-step process with a high-purity DI water rinse following each step. The first step uses a mixture of one part peroxide (H_2O_2), one part ammonium hydroxide (NH_4OH), and five parts DI water. The solution is heated to 80°C. The wafers are immersed in the cleaning solution for 10 - 15 minutes and then are immediately placed in the second solution. The second solution is a mixture of one part peroxide, one part hydrochloric acid (HCl), and six parts DI water. This solution is also heated to 80°C. The wafers are left in this solution for 10 - 15 minutes followed by a DI water rinse and are dried. This process is most often used before a high-temperature processing step.

The two steps are sometimes called the SC-1 and SC-2 cleans where SC-1 is the DI water, peroxide, ammonium hydroxide solution and SC-2 is the DI water, peroxide, hydrochloric acid solution.

The sulfuric acid and hydrogen peroxide solution, often written as H_2SO_4-H_2O_2, is most often used to strip photoresist from the wafer surface.

The sulfuric acid-hydrogen peroxide solution is formulated by heating the sulfuric acid to 80 - 100°C and adding the peroxide just prior to the wafer cleaning step. The solution contains three to five volumes of sulfuric acid to one volume of peroxide. Since both sulfuric acid and hydrogen peroxide are strong oxidizing agents, mixing them together causes an exothermic chemical reaction (liberates heat) raising the solution temperature to 150 - 180°C, depending on the chemical ratios. This sudden increase in temperature aids in the chemical cleaning process. The wafers are immersed in the solution for 10 - 15 minutes followed by a DI water rinse and are dried.

C. DIELECTRIC FORMATION

A dielectric is a material that is a poor conductor of electricity or does not conduct at all. Dielectric layers are often called insulators. Dielectric layers play a critical role in the manufacturing and operation of semiconductors. They are used:

- for insulation between conducting layers (e.g., for devices with more than one level of metal and to separate the gate from the silicon in an MOS transistor),

- to protect the surface of a completed die,

- to mask off portions of the surface of the wafer during some manufacturing operations, and

- between the plates of capacitors in ICs.

Various forms of dielectrics related to semiconductors include silicon dioxide, silicon nitride, and silicon oxynitride. The most common is silicon dioxide.

1. Thermally Grown Silicon Dioxide

Silicon has a very unique property when exposed to any source of oxygen. A chemical reaction that forms silicon dioxide (SiO_2) will occur. Silicon dioxide is a very stable dielectric or insulating material. Silicon will react with oxygen in the air at room temperature to form a very thin layer of silicon dioxide. This layer will be 20 to 30Å thick wherein the thickness stops the chemical reaction. The process of reacting the silicon wafer with some source of oxygen is referred to as thermal oxidation. In other words, silicon at the interface between the oxide formed and the silicon wafer is being consumed by the chemical reaction of forming silicon dioxide. These silicon atoms are permanently changed into the compound, silicon dioxide. The effects of thermal oxidation are shown in Figure 2-9.

Basic Integrated Circuit Manufacturing

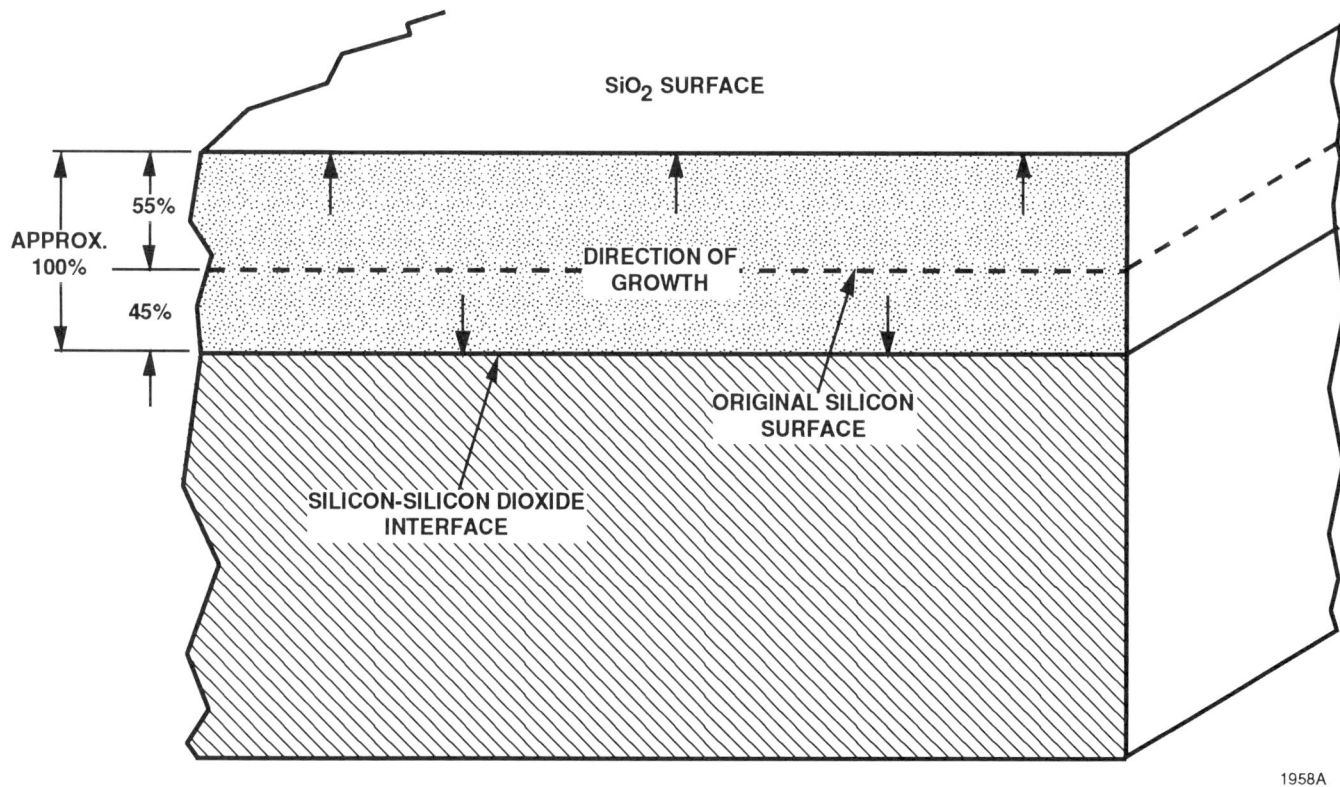

Figure 2-9. The Effects of SiO$_2$ Growth

a. Atmospheric-Pressure Oxidation

In semiconductor manufacturing, thermal oxidation is a well understood process. The process is done in the temperature range of 750°C to 1150°C to increase the reaction rate. The rate of oxidation also depends on the source of oxygen and the pressure inside the process chamber. The process parameters of time, temperature, oxygen source and pressure are determined by the overall process requirements and circuit design considerations.

The thermal oxidation oxidation rate is affected by the process engineer's choice of dry oxygen (pure O$_2$) or some form of water (H$_2$O) vapor. Generally, if a thin oxide (<1000Å) is required, dry oxygen is used. Thicker oxides use water vapor. Oxidation using water vapor formed from the chemical reaction of gaseous oxygen (O$_2$) and hydrogen (H$_2$) is known as pyrogenic steam oxidation. This process has replaced the older process of using deionized water. The previous types of thermal oxidation are done at atmospheric pressure and illustrated in Figure 2-10.

Basic Integrated Circuit Manufacturing

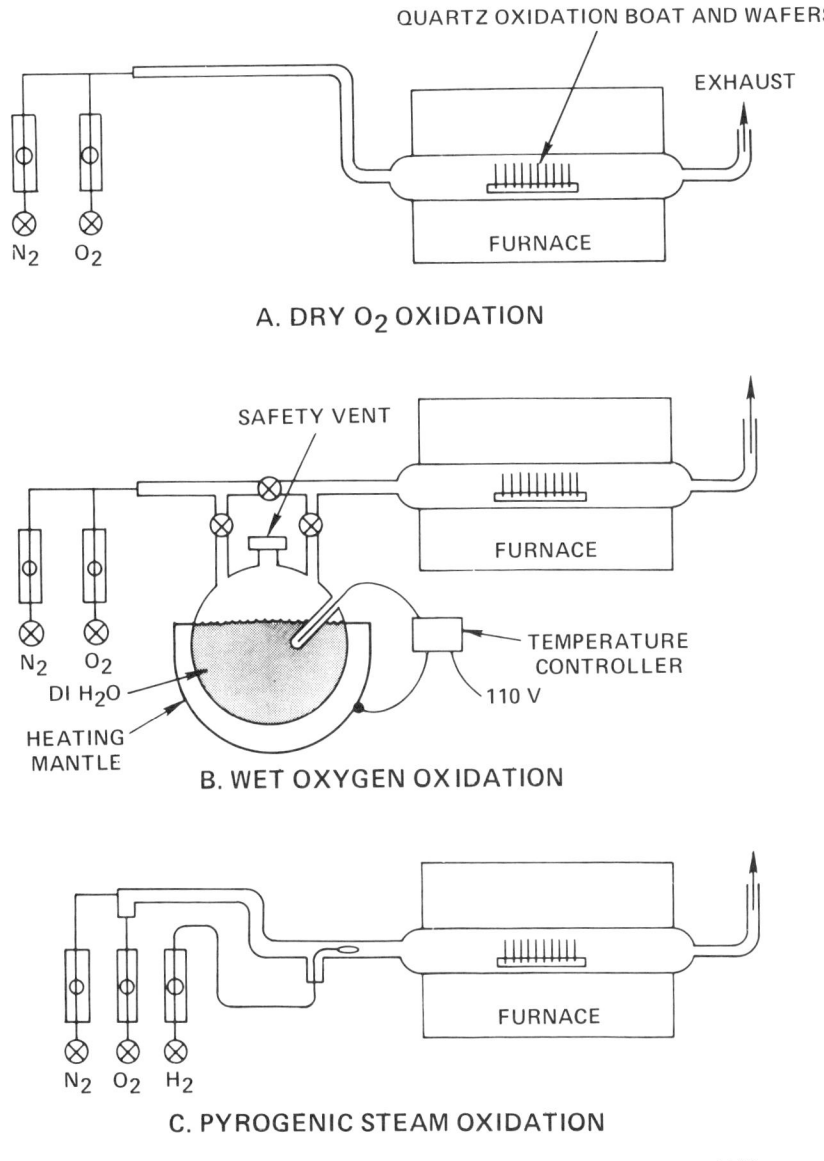

Figure 2-10. Oxidation Systems

b. High-Pressure Oxidation

As feature sizes of integrated circuits continue to decrease further into the submicron size, more shallow junctions are required and process temperatures are therefore being forced to lower levels. This has brought about a renewed use of high-pressure oxidation systems. These systems originated in about 1970, but have been slow to evolve into manufacturing.

The process chamber on this system is fitted with a High-Pressure Vessel. This allows the system to be pressurized from one to twenty-five (25) atmospheres (14.7 psi to 367.5 psi). This allows the process engineer to lower temperature by raising the pressure of the process chamber and has allowed some new degrees of process choices. In addition, this equipment has shown a reduction in oxidation defects when compared to the atmospheric-pressure process.

The disadvantages of high-pressure oxidation are the higher cost of the equipment, the equipment uses twice the floor space, and the safety considerations are stringent.

The advantages today more than offset the disadvantages because of the need for lower temperature processing.

2. CVD Silicon Dioxide

Silicon dioxide can be formed by manufacturing techniques other than thermal oxidation. One is known as Chemical Vapor Deposition (CVD). This process is done within a confined volume or separate process chamber. The process reactions bring about a chemical decomposition of certain elements or compounds by using heat, light, pressure/vacuum, and/or plasma energy to form a stable solid.

There is a clear distinction to be made between thermal oxidation to form silicon dioxide and CVD depositions to form silicon dioxide. Thermal oxidation reacts some form of oxygen with the silicon wafer at elevated temperatures. Thus, the silicon wafer is the source of silicon. When silicon dioxide is formed from a CVD process, both the silicon and the oxygen are brought to the process chamber from external sources. Thus the silicon wafer *is not* part of the chemical reaction. The silicon wafer is coated with the results of the CVD chemical reaction, silicon dioxide. Because of this difference, each type of silicon dioxide, thermal and deposited, has a unique set of physical characteristics. The characteristics determine which type of oxide is used at a given manufacturing step.

There are many silicon compounds available for use in the CVD process. However, silane (SiH_4) and dichlorosilane (SiH_2Cl_2) are the more commonly used materials. Oxygen for CVD can be pure dry oxygen, or can come from decomposing some compound of oxygen. Commonly used compounds are carbon dioxide (CO_2), nitrous oxide (N_2O), and nitrogen dioxide (NO_2).

a. Uses of CVD-Deposited Silicon Dioxide

CVD-deposited oxides are used at several different places in the manufacturing sequence. The following lists some of the more common uses.

1. To increase oxide thickness of thermal oxides
2. Capacitor dielectric
3. Dielectric over polysilicon

4. Dielectric over metal
5. Buffer oxide layer to match mechanical requirements
6. Masking oxide layer
7. Final passivation

Deposited oxide is not normally used as a gate oxide for MOS transistors. Gate oxide is formed by some thermal oxidation technique.

b. CVD Equipment

The physical hardware for CVD-deposited oxide can be configured many different ways. CVD reactors can be classified by energy source, pressure, and chamber design. This is illustrated in Figure 2-11. The primary purpose of this process is to deposit a uniform film of silicon dioxide that has a certain composition. The LPCVD (Low-Pressure CVD) option is becoming the hardware of choice to meet this requirement. The system is shown in Figure 2-12.

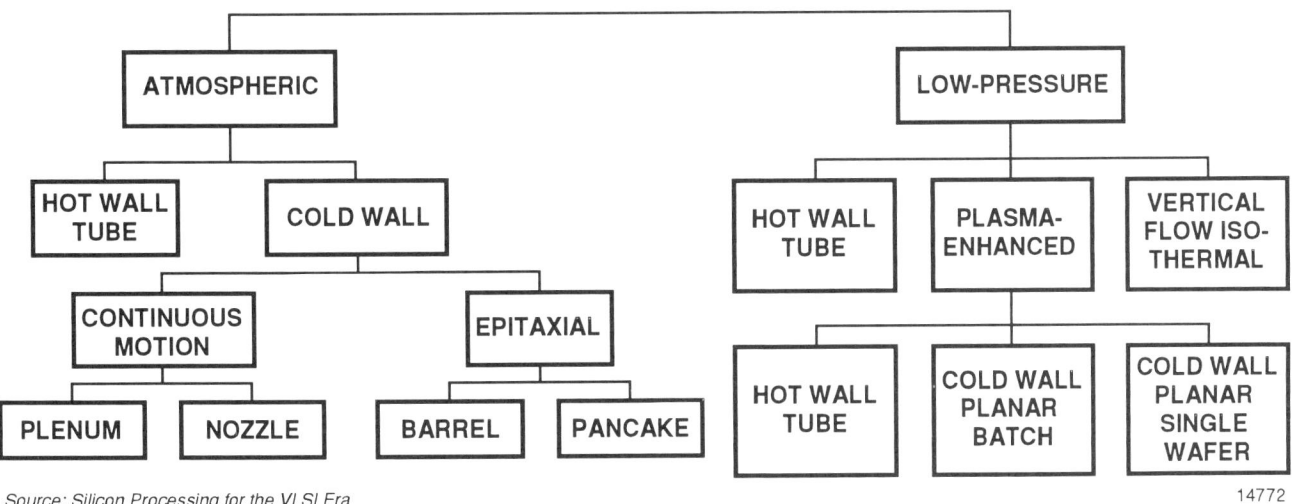

Source: Silicon Processing for the VLSI Era

Figure 2-11. CVD Reactor Types

It has been found useful for some deposited silicon-dioxide films to have other elements added during the deposition. Two of the more common elements added are phosphorus and boron. These elements can be added individually to the silicon dioxide and sometimes they are both included in the silicon-dioxide film at the same time during the deposition process.

In addition to silicon-dioxide films deposited by CVD technology, silicon nitride (Si_3N_4), silicon-oxynitride ($Si_xO_yN_z$), and various forms of polysilicon are deposited. Each of these films has certain chemistry considerations along with temperature and pressure requirements. Typical reactions for CVD are tabulated in Figure 2-13.

Basic Integrated Circuit Manufacturing

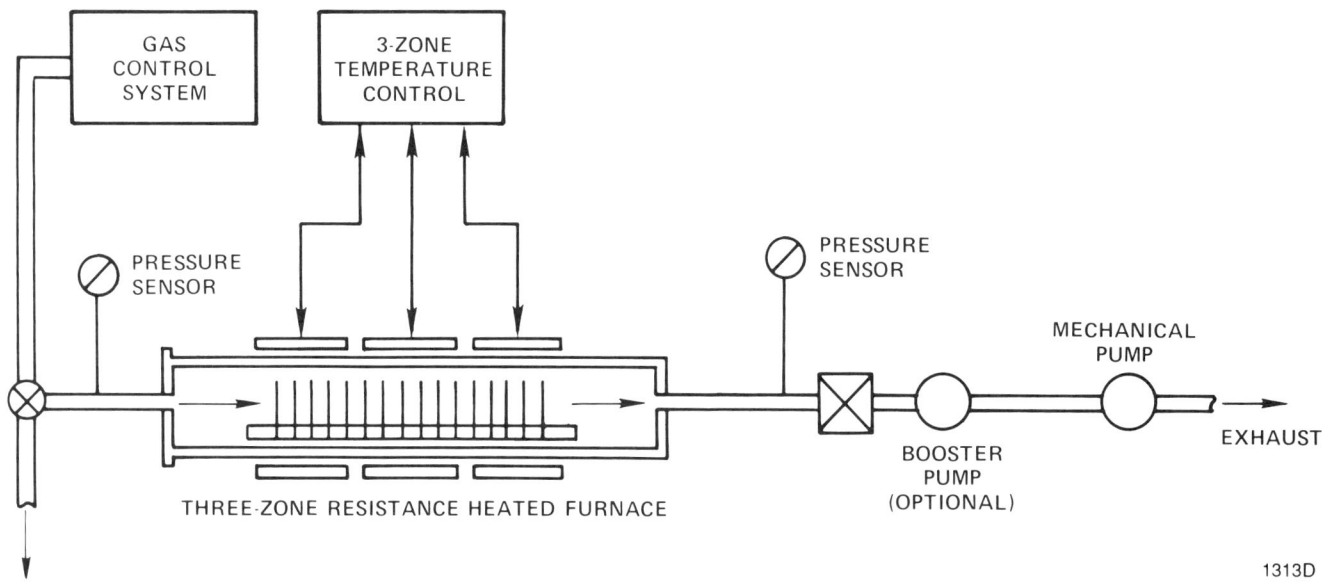

Figure 2-12. Block Diagram of a Low-Pressure Chemical Vapor Deposition System

PRODUCT	REACTANTS	DEPOSITION TEMPERATURE (°C)
SILICON DIOXIDE	$SiH_4 + CO_2 + H_2$ $SiCl_2H_2 + N_2O$ $SiH_4 + N_2O$ $SiH_4 + NO_2$ $Si(OC_2H_5)_4$ $SiH_4 + O_2$	850 – 950 850 – 900 750 – 850 650 – 750 650 – 750 400 – 450
SILICON NITRIDE	$SiH_4 + NH_3$ $SiCl_2H_2 + NH_3$	700 – 900 650 – 750
PLASMA SILICON NITRIDE	$SiH_4 + NH_3$ $SiH_4 + N_2$	200 – 350 200 – 350
PLASMA SILICON DIOXIDE	$SiH_4 + N_2O$	200 – 350
POLYSILICON	SiH_4	600 – 650

Source: VLSI Technology

Figure 2-13. Typical Reactions for CVD Depositions

D. PHOTOLITHOGRAPHY

1. Overview

In the photolithography process sequence, the wafer is covered with a layer of light-sensitive material (photoresist), which is then selectively exposed to light. The selective exposure is accomplished by shining the light through a quartz plate (mask or reticle) with a patterned opaque material on it (Figure 2-14).

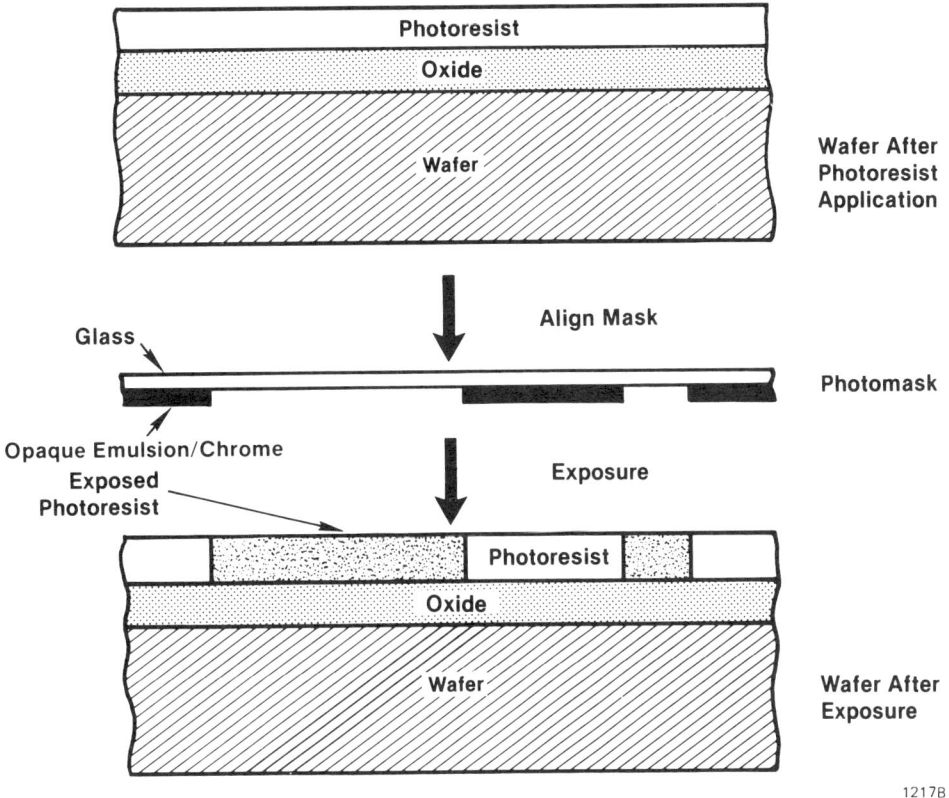

Figure 2-14. Alignment and Exposure

The exposed photoresist is washed away and the remaining, unexposed photoresist is hardened by baking. The portions of the layer below the photoresist not covered by the hardened photoresist is removed and then the photoresist is removed (Figure 2-15).

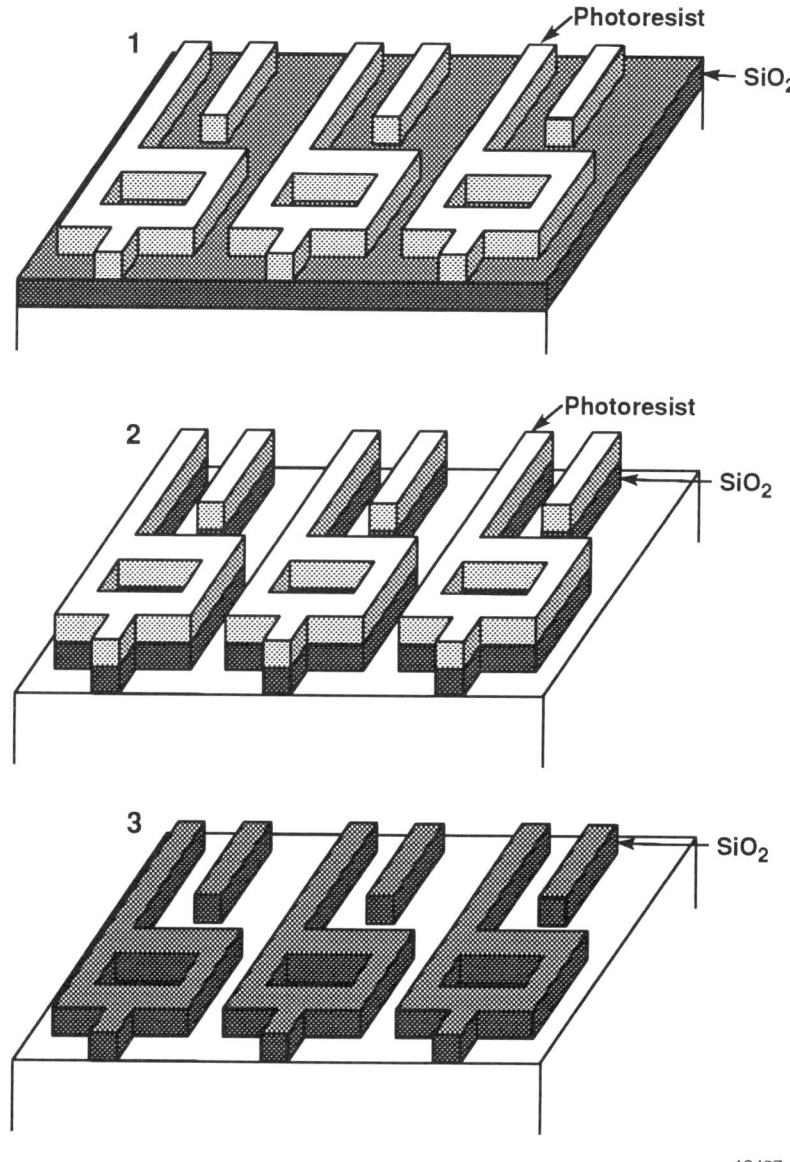

Figure 2-15. Photolithography Using Positive Photoresist

The following is an in-depth analysis of photolithography.

2. Introduction

The photolithography process encompasses all of the patterning operations necessary to transfer an image from one medium to another. The term, photolithography, implies the use of light as part of the transfer process. Frequently the term, microlithography, is used to mean the photolithography process for transferring feature sizes in the submicron range.

Basic Integrated Circuit Manufacturing

The photolithography process involves the transfer of geometric images created by a circuit designer to a photosensitive film applied to the surface of the silicon wafer. The sized geometrical features are the physical designs of the various circuit elements, i.e., transistors, resistors, capacitors, etc., that will make up the electrical circuit design.

Since the early 1970's the concept of making the physical size of the transistor smaller to increase the transistor performance has been the evolutionary force behind the semiconductor industry. This has placed a constant pressure on the photolithography process to print smaller feature sizes.

Figure 2-16 illustrates the critical aspects of photolithography. There are many patterns that must be transferred to the wafer, each sized properly and registered (aligned) to one another in the correct sequence. The transfer starts with the circuit designer creating the electrical circuit, aided by the power of the computer. After the electrical circuit is created, the circuit schematic is converted into the physical sizes and shapes of the circuit elements that are arranged (layout) in a surface plane and the electrical connections are made to each circuit element.

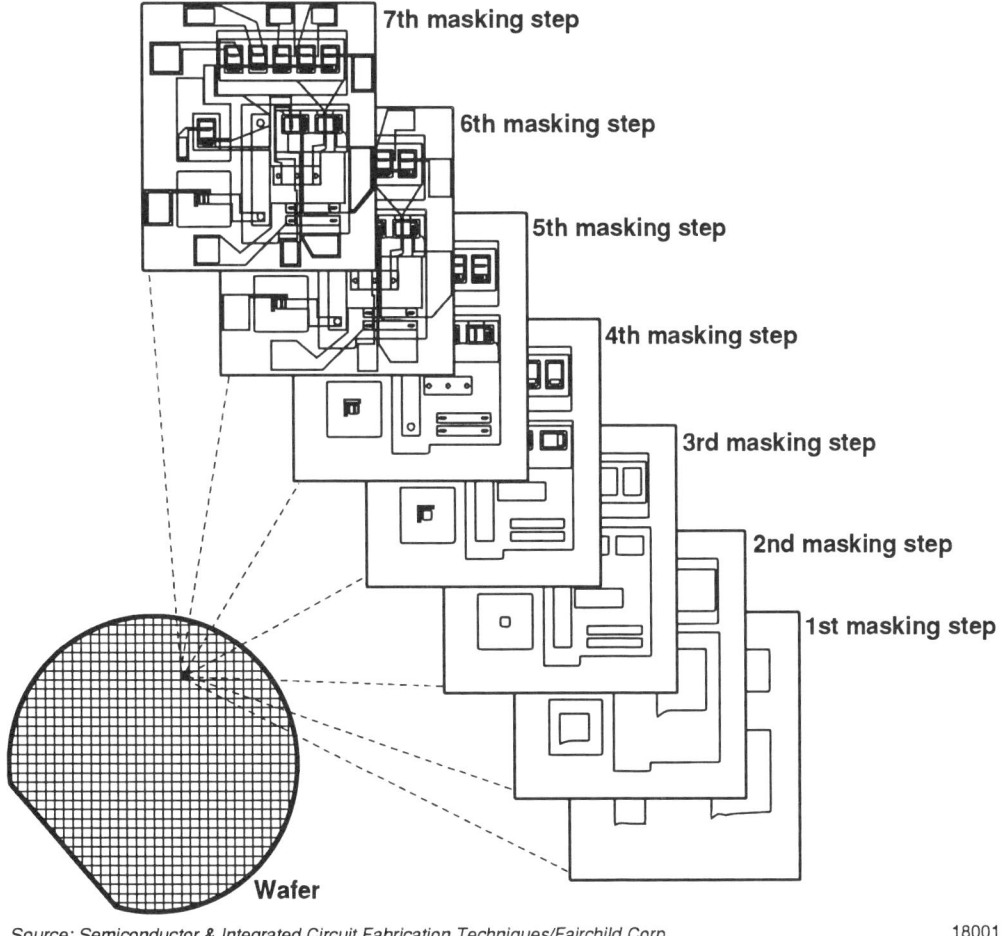

Source: Semiconductor & Integrated Circuit Fabrication Techniques/Fairchild Corp. 18001

Figure 2-16. The Layers Transferred to a Wafer During a Seven-Mask Process

INTEGRATED CIRCUIT ENGINEERING CORPORATION

3. Masks

A photomask is a quartz plate with one layer of patterns for all of the ICs on a wafer on it (Figure 2-17). A reticle only has the patterns for a few ICs on it (Figure 2-18). The patterns are formed with opaque substances such as emulsion or chrome.

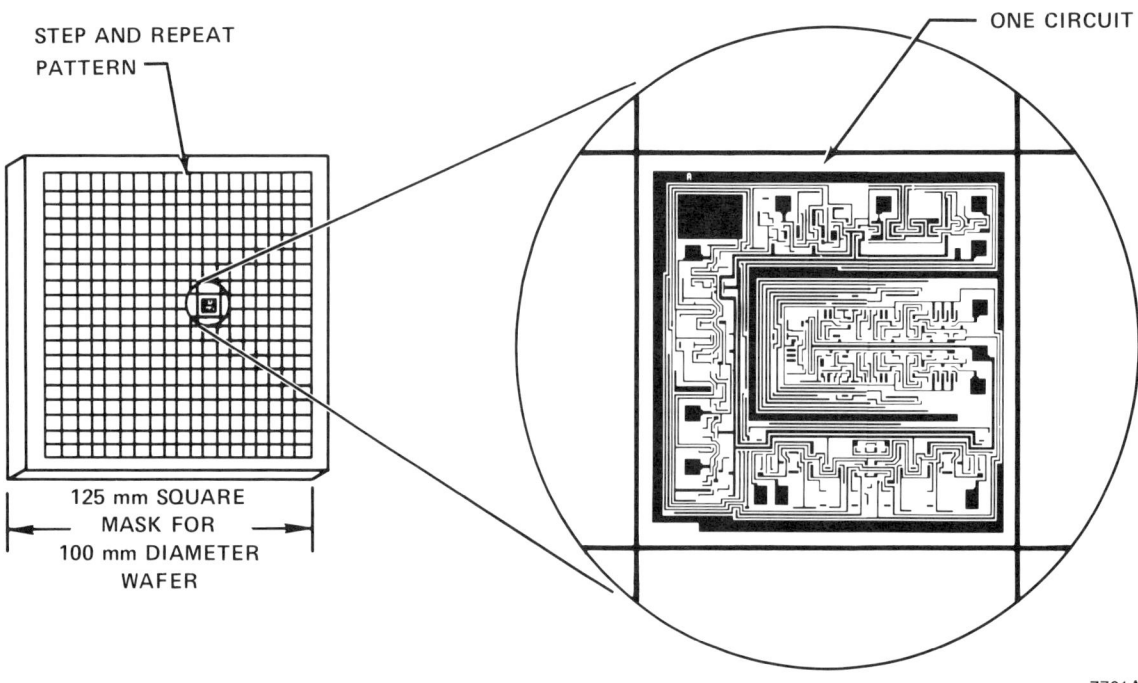

Figure 2-17. Example of a Photomask

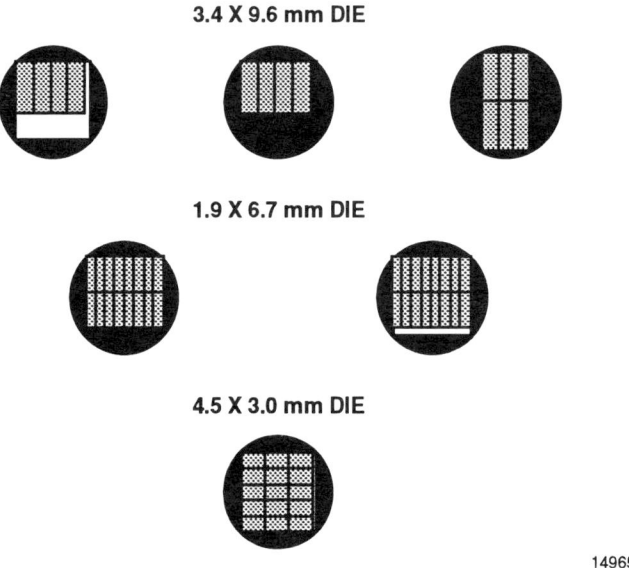

Figure 2-18. Die Placement in Fixed Reticle of 21mm

Emulsion images on the glass substrate are difficult to clean. Typically they are used for a limited number of exposures and then discarded. On the other hand, chrome images on the glass substrate can be cleaned, inspected, and reused many times.

a. Reticles

The design process is illustrated in Figure 2-19. The information generated from the design process is used to manufacture a reticle for each layer of the circuit. The reticle is an intermediate-sized representation of the circuit used in the image transfer process. The reticle can be sized as large as ten times (10X) the actual circuit size down to the actual circuit size (1X).

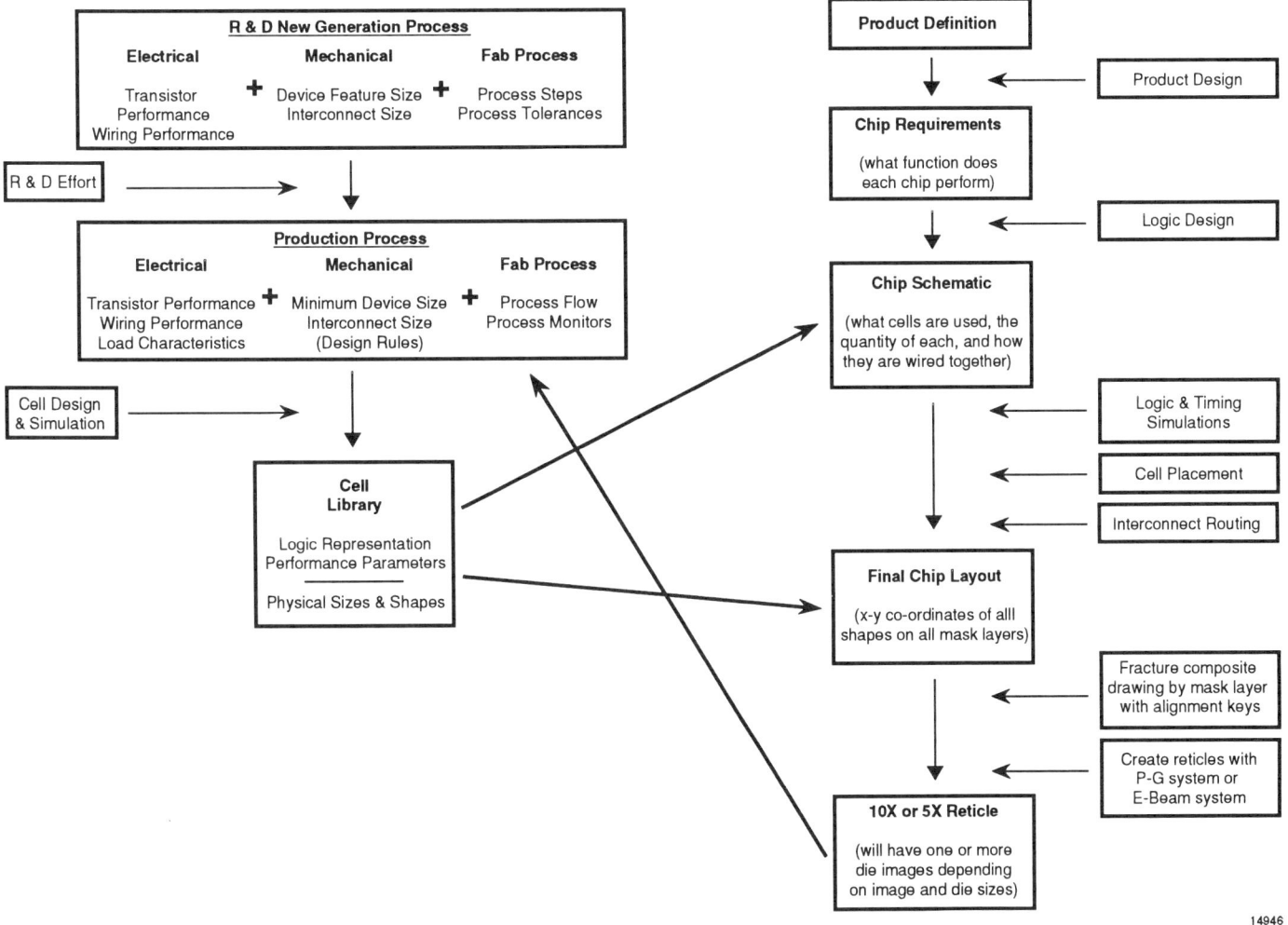

Figure 2-19. The Design Process

Basic Integrated Circuit Manufacturing

The manufacturing procedures and equipment used for reticle generation require the best precision and reproducible imaging technology known. The quality of the reticle can impact the yields of the actual IC manufacturing.

The manufacturing of the reticle can follow a variety of paths. The alternative routes for reticles' manufacturing are shown in Figure 2-20.

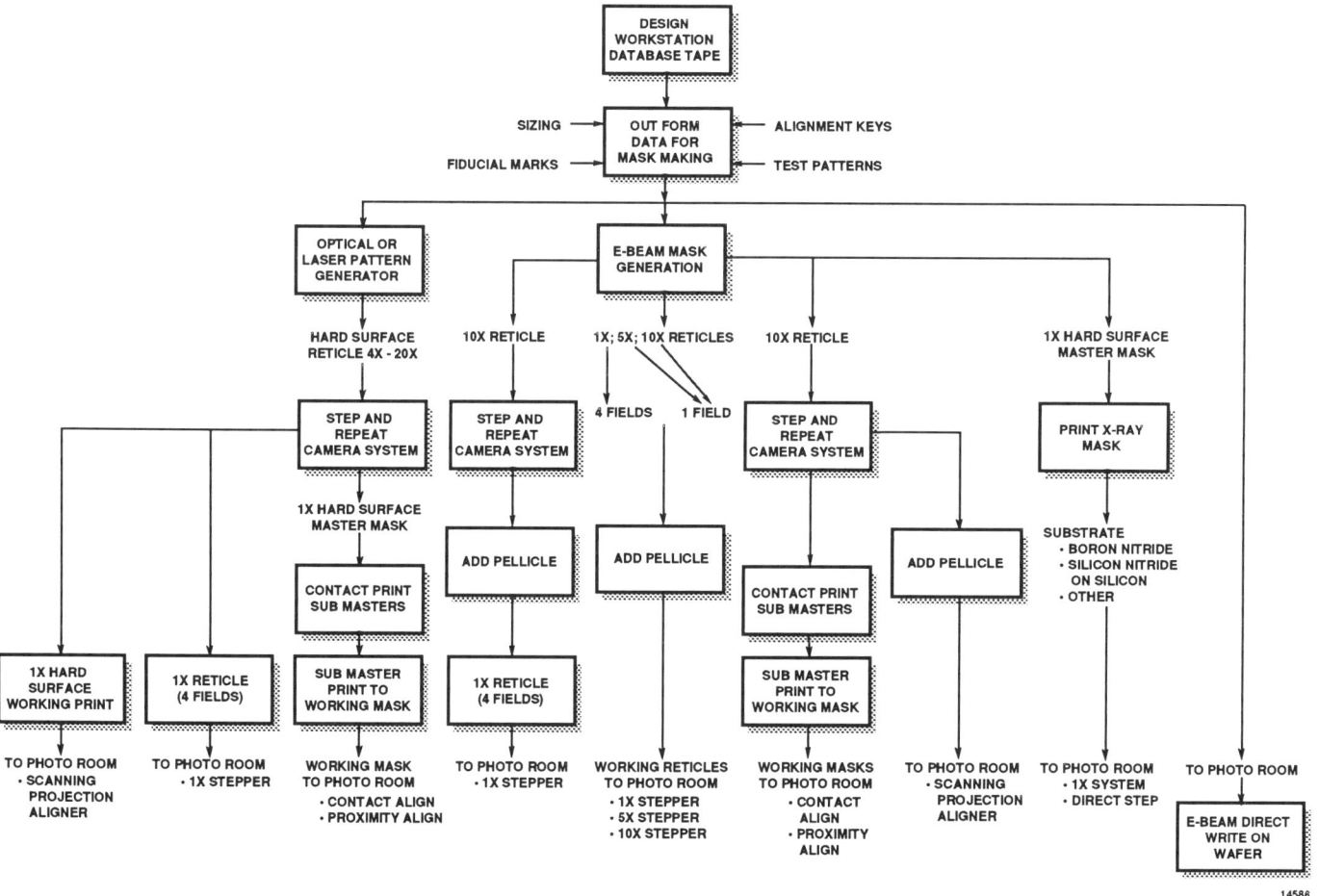

Figure 2-20. Microlithography Roadmap

Reticles for Direct Step on Wafer are made in 10X, 5X, 4X, or 1X size, depending on the type of stepper equipment chosen. A reticle's size is determined by the magnification of the stepper lens and usable field size of the stepper lens. This will determine how many dice can fit within the usable field area. Figure 2-18 illustrates different die sizes fitting into a 21mm field size.

b. Photomasks

The photomask is created by a series of additional process steps beyond the making of a 10X reticle for each layer. Refer to Figure 2-20. At the contact print to submaster stage the mask polarity must be determined (depends on whether negative or positive photoresist is used). The submaster to working print stage can print either emulsion or chrome on glass substrates. The choice is determined by the wafer alignment equipment. A photomask is shown in Figure 2-17.

c. E-beam

Also shown in Figure 2-20 are two other alternative technologies to form the designer's requirements into a photoresist layer on the silicon wafer. One alternative is to use an e-beam system to write the pattern directly on the photoresist-coated wafer. This approach eliminates the reticles or mask-making stage but has a very slow throughput.

d. X-ray

The other technique uses x-ray energy to expose the photoresist-coated wafer. The mask for this technique is very expensive and difficult to manufacture. However, x-ray steppers have excellent resolution capability and will probably be needed by the end of the 1990's.

4. Photolithography Sequence

Figure 2-21 illustrates the manufacturing sequence for photolithography. These photolithographic steps encompass all of the patterning operations, including wafer priming, coating, align, expose, develop, etch, and photoresist removal. The characteristics of the light-sensitive photoresists determine the basic process technique. As was shown in Figure 2-16, the photolithographic process is repeated several times. Because of this it is often referred to as the hub of the IC fabrication process.

a. Photoresist, Negative and Positive

There are two kinds of photoresist commonly used: negative and positive. The chemical behavior of each resist is illustrated in Figure 2-22. The negative resist responds to the radiation (UV light) in a manner that prevents the developer solution from removing the exposed resist. The image formed in the resist is the same as the clear area on the mask. The unexposed resist is removed by the developing process. Positive photoresist has the opposite response to the radiation. The areas of the photoresist that are exposed are removed by the developer solution. Thus, the unexposed resist remains and forms the image on the surface of the wafer.

The chemistries of positive and negative photoresist are very different. Positive photoresist is developed with a mild alkaline (basic) solution and the negative resist requires a solvent (xylene) for developing.

Basic Integrated Circuit Manufacturing

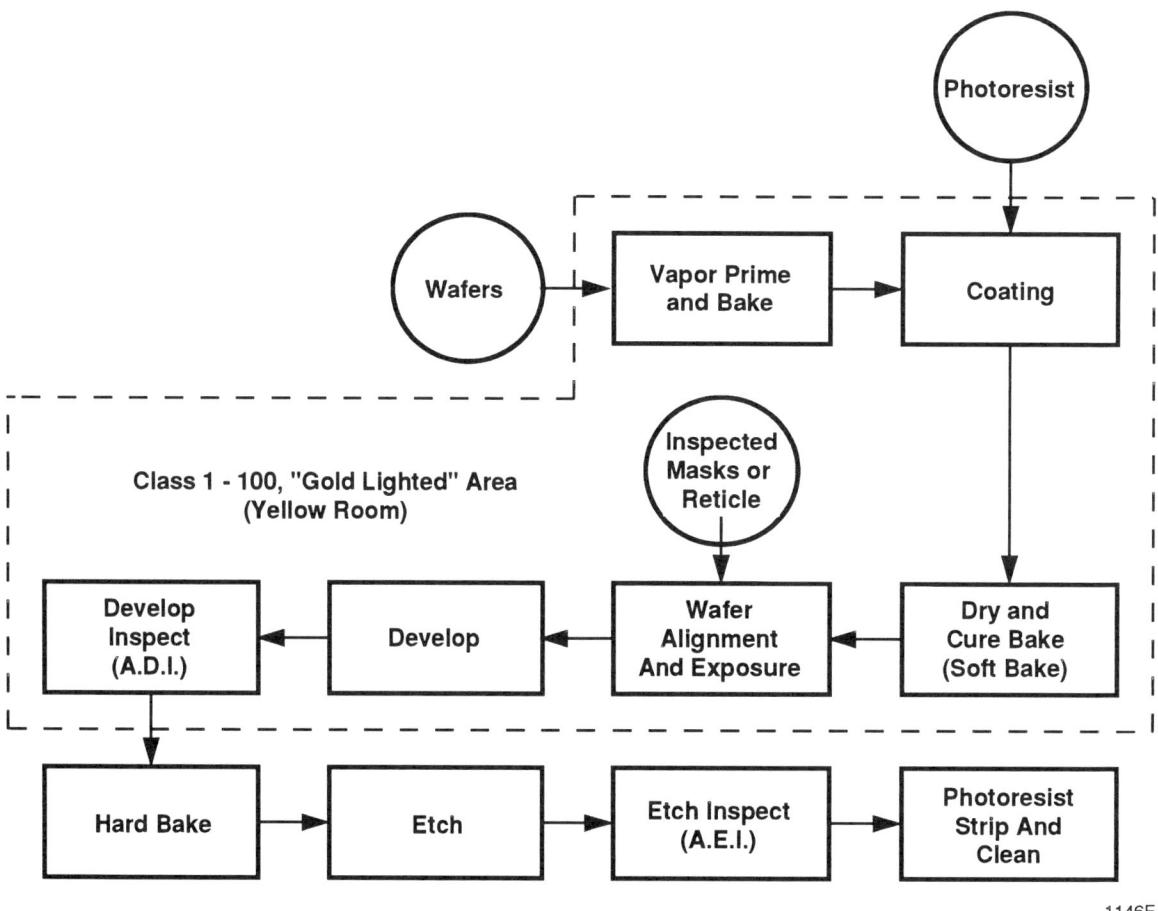

Figure 2-21. Photolithography Process Flow Chart

i. Exposure Wavelengths

A portion of the electromagnetic spectrum is illustrated in Figure 2-23. The blue-violet region is referred to as the ultraviolet wavelengths. Also shown is the relationship between wavelength (λ) and the frequency (f) for this form of energy.

The light-sensitive responses of both resists are similar in that both are exposed by light in the blue-violet wavelength (190 - 450nm). These wavelengths are commonly found in mercury arc lamps and similar bulbs. For this reason, the process area for photoresist must have these wavelengths filtered out during manufacturing to prevent unwanted exposure. Yellow filters or lights are used to illuminate the work area for these processing steps.

Basic Integrated Circuit Manufacturing

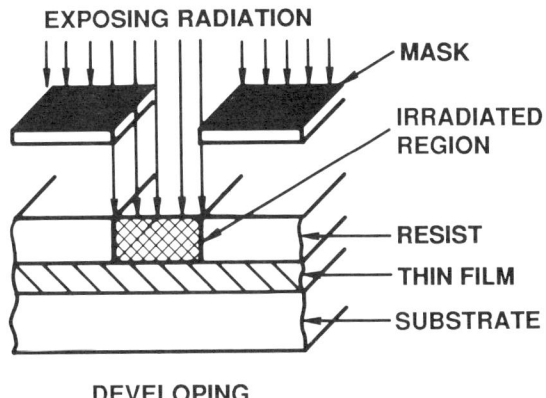
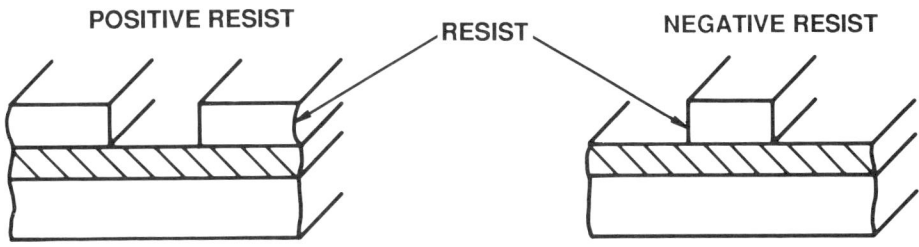

Figure 2-22. Characteristics of Negative and Positive Resists

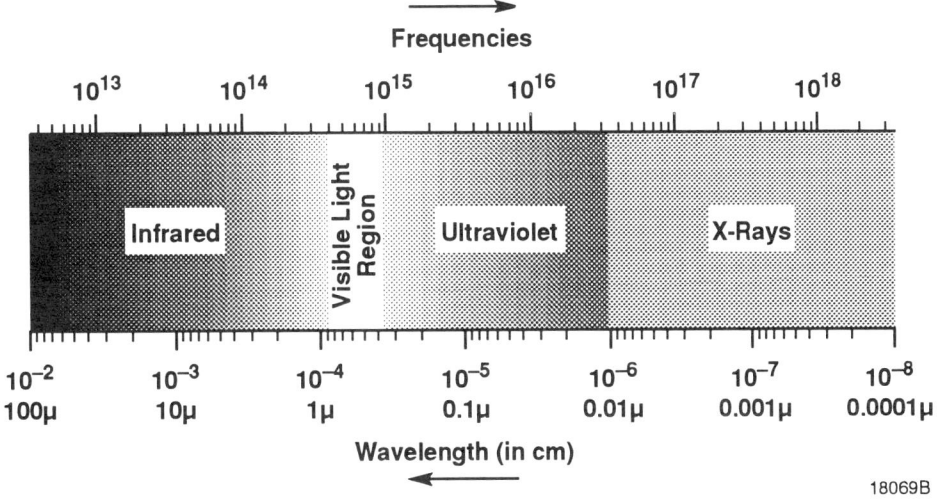

Figure 2-23. Part of the Electromagnetic Spectrum

ii. Photoresist Parameters

Photoresist is typically characterized by several parameters that affect its performance. These are:

1. adhesion
2. etch resistance
3. resolution
4. photo sensitivity
5. step coverage

b. Wafer Preparation Before Photoresist Application

i. Cleaning

The photoresist process is extremely sensitive to any form of contamination on the surface of the wafer prior to applying the photoresist. To assure the best surface possible, various cleaning techniques are often used prior to the priming process. The cleaning process may be some type of wafer scrubbing (brush or high pressure), chemical cleaning with agitation, or chemical cleaning only.

ii. Priming

The surface of the silicon substrate can have various layers depending on where in the manufacturing cycle the current operation resides. Each layer on the silicon surface has a different effect on photoresist adhesion. To provide better adhesion, a process step called priming is performed. The use of a priming solution increases the adhesion of the photoresist to the surface. Primers may be applied by immersing the substrates in the priming solution, spraying the solution on, or by passing a priming vapor over the surface of the wafer. Some primers have to be baked before subsequently coating the wafers with photoresist. Other priming techniques require that the wafer surface be dehydrated at elevated temperatures prior to applying the priming solution.

c. Photoresist Application

Photoresist may be applied to the surface of the wafer using a variety of techniques. These techniques include dipping, spraying, brushing, roller coating, and spin coating. Spin coating is the method most often used in the fabrication of semiconductors.

The spin coating process applies a resist layer to the silicon substrate as uniform as possible for the required thickness. The uniformity of the coating is very important. As the wafers progress further through the manufacturing cycle, the topology of the surface continues to change in thickness above the silicon surface. The variation in the vertical heights causes some variation in the resist thickness uniformity.

The coating process involves holding the wafer on the motor spindle by vacuum. The resist is dispensed onto the wafer surface. The wafer is slowly spun at a low RPM to spread the resist over the entire wafer surface. After a few seconds of low RPM, the spindle will rapidly accelerate to a much higher RPM (4,000 to 6,000) for final thickness control. The wafer is decelerated back to a low RPM wherein a resist thinner solution is dispensed around the outer 2mm of the edge of the wafer to remove the bead of resist that builds up from the spinning process. Removing this bead eliminates a source of contamination from the process.

The wafer is transferred to a soft bake heat source to remove the solvent from the resist. This leaves a firm resist layer ready for the align and expose process. A cross section of the structure is illustrated in Figure 2-24.

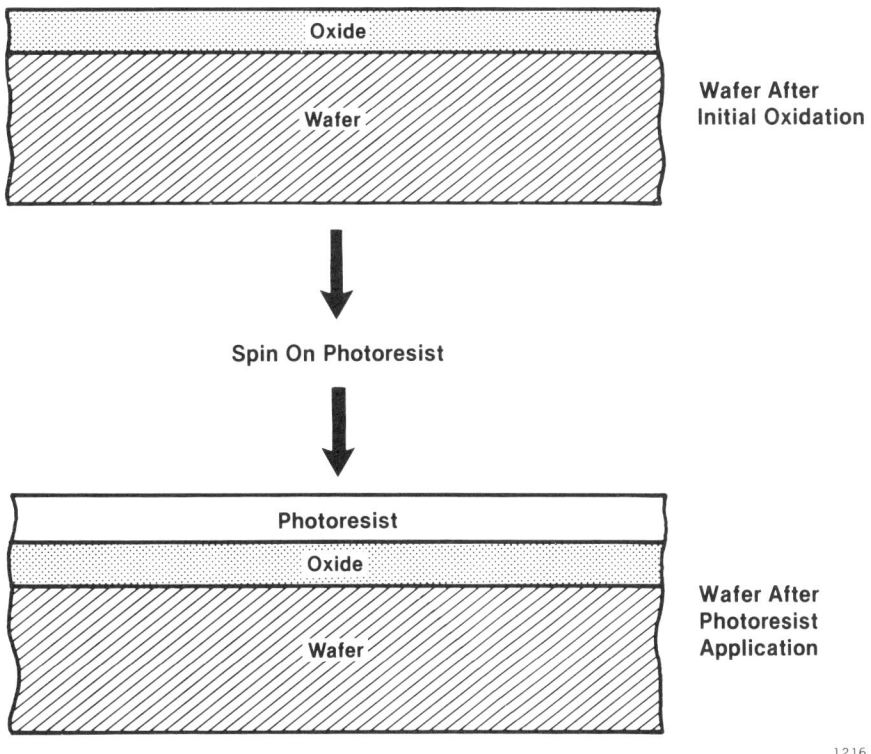

Figure 2-24. Photoresist Application

d. Alignment/Exposure

The process of forming an image in the photoresist-coated surface has undergone considerable change. Early wafer align/expose systems typically worked on the principle of "global" or "blanket" exposure, in which the entire wafer was exposed at once. As resolution requirements kept pushing the feature size smaller and smaller and overlay registration (lining up one layer to preceding layers) had to become more accurate, alternatives were investigated.

Basic Integrated Circuit Manufacturing

The align/expose process was illustrated in Figure 2-14. There are several types of align and expose equipment (Figure 2-25). The oldest alignment system is the contact aligner and is normally used in older fabs manufacturing older products. The contact printer concept is illustrated in Figure 2-26.

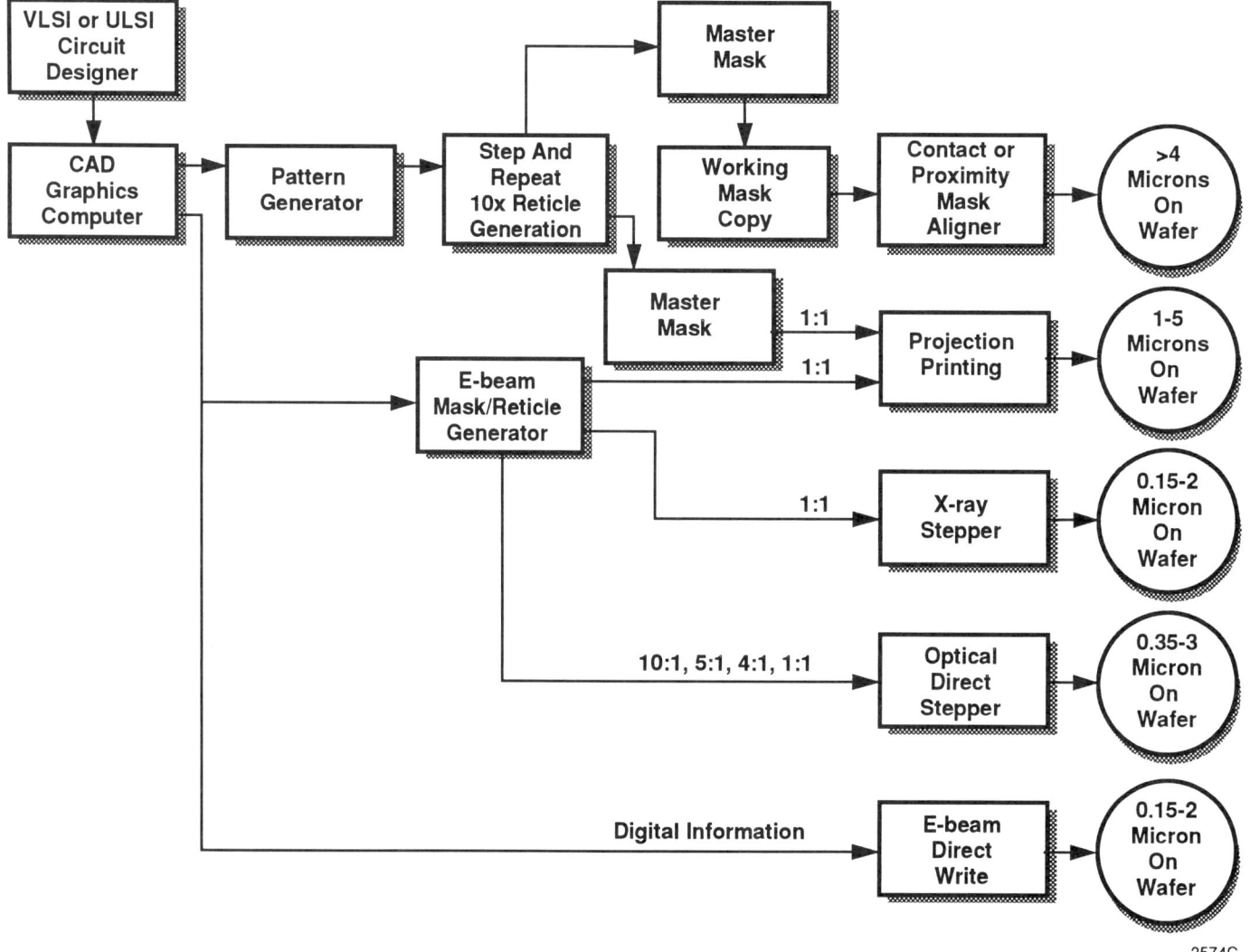

Figure 2-25. Wafer Patterning Systems

Basic Integrated Circuit Manufacturing

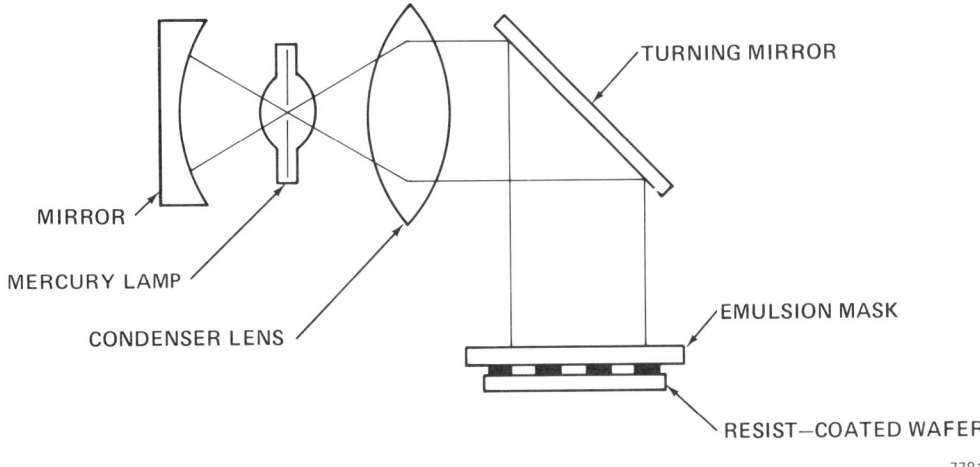

Figure 2-26. Contact Printer Optical Configuration

i. Contact Aligner

The contact aligner can achieve very good resolution but has poor overlay registration and a high defect density. The high defect density is the result of the mask and wafer making contact during the exposure. This contact between the mask and wafer can create physical defects in both the mask and wafer. As the mask is used to align and expose additional wafers, the defect density will continue to increase. The detrimental effect is shown in Figure 2-27.

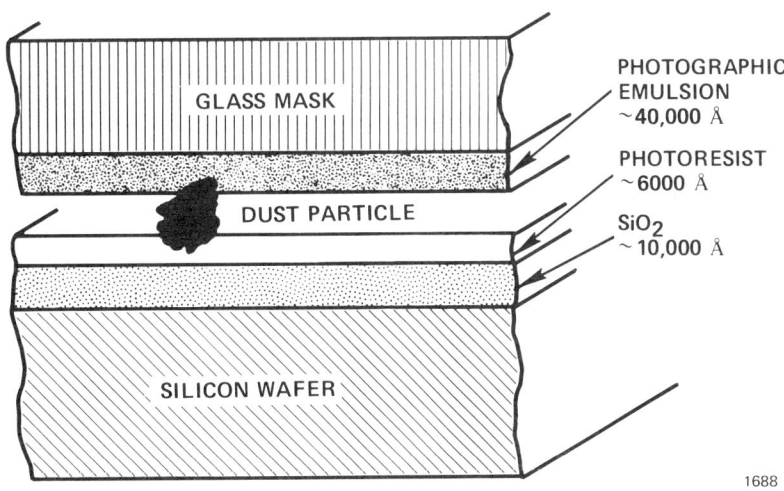

Figure 2-27. Dust Particle Scratching Photoresist and Photomask

ii. Proximity Printing

An early alternative to contact printing was proximity printing. The light source in this equipment is colliminated enough to allow a 10 to 20um separation between the mask and the photoresist surface. This equipment concept is illustrated in Figure 2-28. An improvement in defect density was realized but the resolution was typically limited to 4μm or greater.

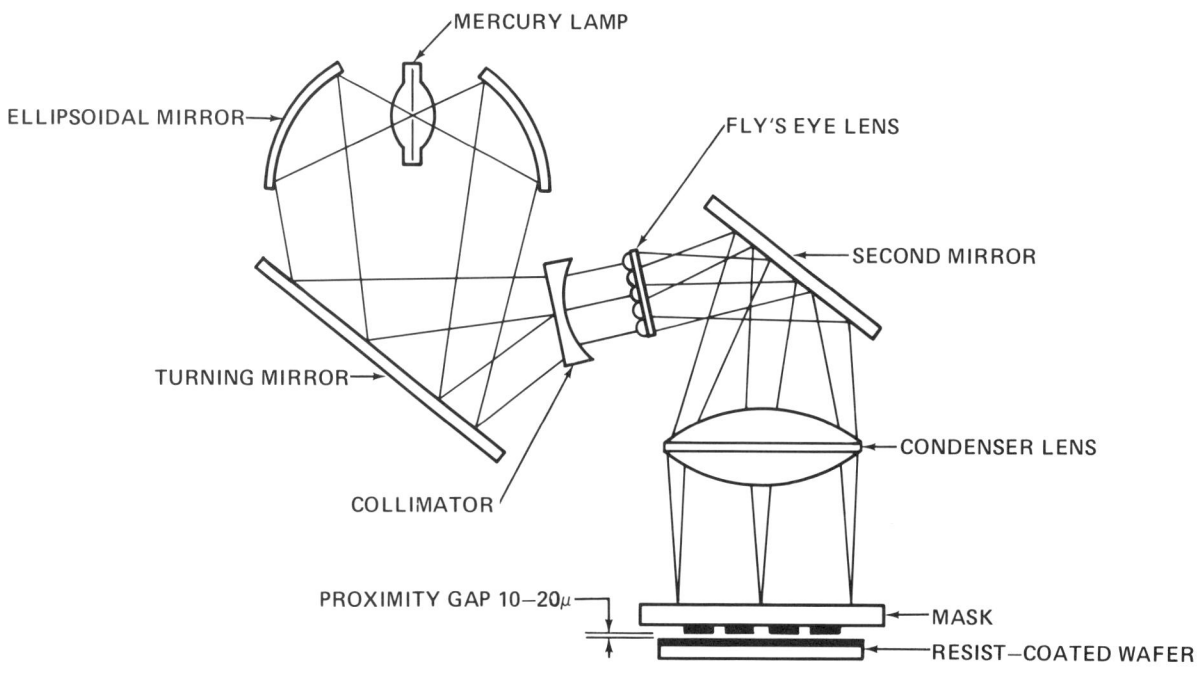

Figure 2-28. Proximity Printer Optical Configuration

iii. Projection Alignment

The next approach to improving the align/expose process was the introduction of the projection aligner. In this system there is no contact between the wafer and the mask. Rather than blanket exposure, where the entire mask is exposed with a flood of light, the exposure is accomplished with an arc of light. The wafer and the mask move through the equipment coincidentally on parallel planes. The system is illustrated in Figure 2-29.

The projection aligner needs better temperature and humidity control, a cleaner cleanroom, and reduced vibration in the photo room. The aligner also places more stringent requirements on the mask and photo process to achieve an improved defect level.

Basic Integrated Circuit Manufacturing

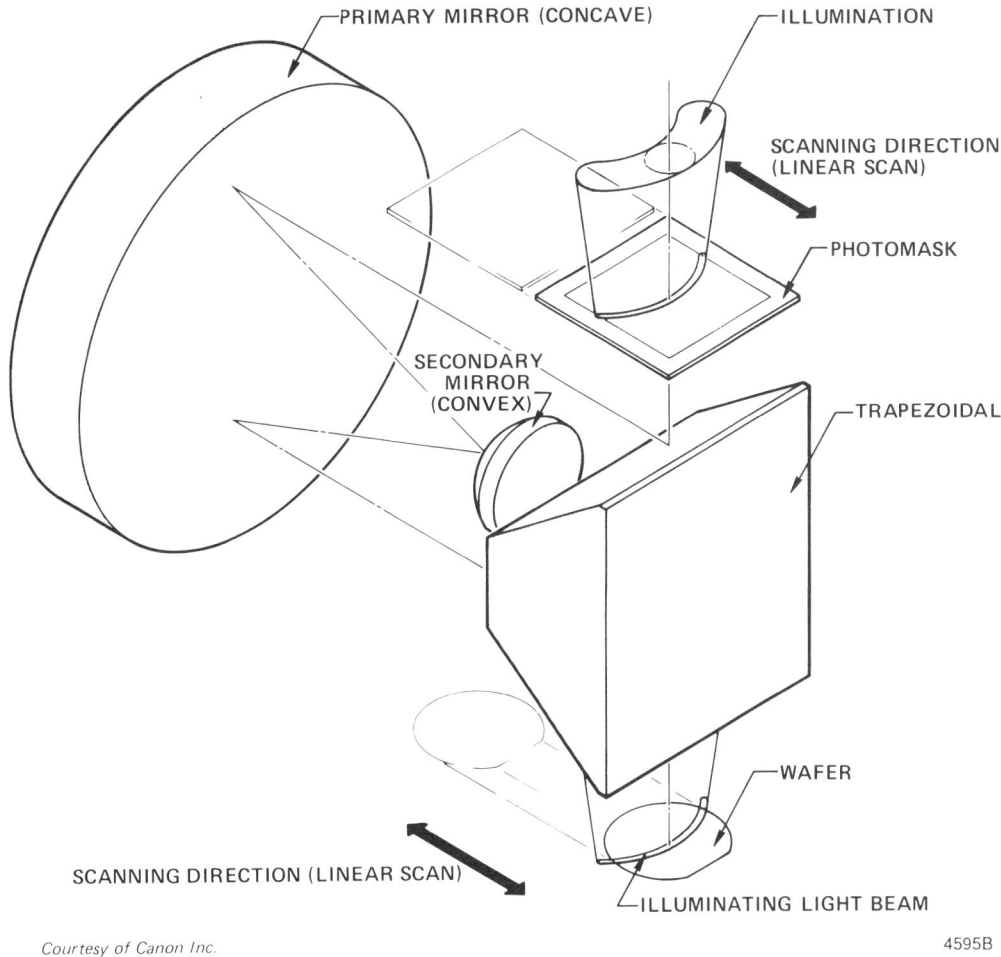

Figure 2-29. Principle of Scanning Projection Aligner

Over several generations of equipment, the projection aligner has become one of the most widely used alignment systems in IC manufacturing. The equipment capability has been extended to slightly below one micron.

With the concept of "mix and match" alignment systems, the projection aligner will continue to be used for several more years as long as the overlay registration can keep pace with the circuit designers' requirements.

iv. Direct Step on Wafer

As feature size progressed below the three-micron size, another align/expose system was introduced to IC manufacturing. This system is known as the Direct Step on Wafer (DSW or Stepper).

When originally introduced, the system used the 10X reticle for the masking process. This eliminated the requirement for the global mask.

The DSW system ushered in a new technology that offered exciting new possibilities for both resolution and overlay registration. Taking advantage of the precision developed for mask making and with the working image confined to a small place in the center of lens, optical distortion can be minimized. This tremendous gain in optical performance came at the expense of lower throughput. The DSW is illustrated in Figure 2-30.

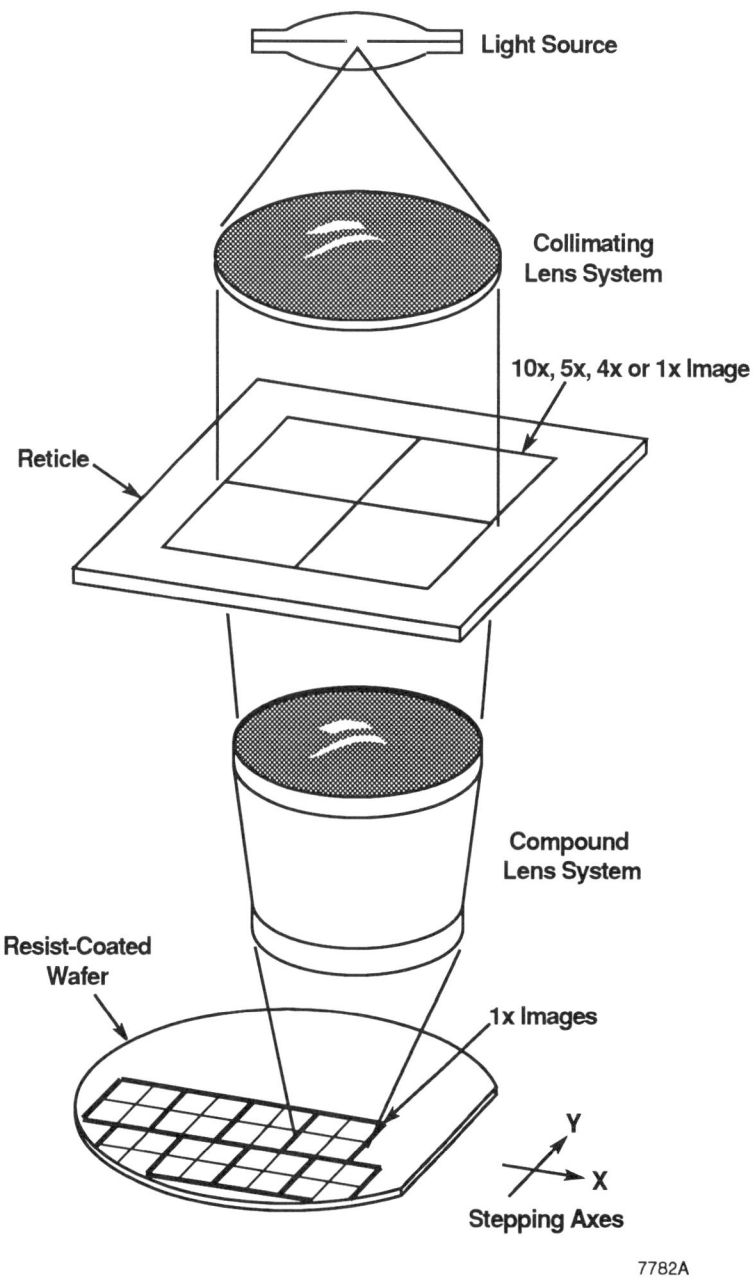

Figure 2-30. Direct Step on Wafer (DSW) Aligner Optical Configuration

Basic Integrated Circuit Manufacturing

- **Mix and Match**

DSW has rapidly gained acceptance but the disadvantage of lower production rates has required larger capital investments. To balance between these two conflicting requirements, the concept of "mix and match" evolved as a reasonable compromise.

In mix-and-match lithography, the scanning projection aligner is used for the less critical levels, and the DSW systems are designated for the more critical mask layers. The net result is better overall yield and a more cost effective capital investment.

To further address the throughput issue, stepper technology found alternative solutions to throughput by reducing the lens magnification. This led to the 5X, 4X and 1X DSW technology. Each of these magnifications has its own compromise but at a slightly improved throughput rate as they decline from 10 to 1.

The issue of lens size vs. die size was illustrated in Figure 2-18. The direct step sequence is illustrated in Figure 2-31 and a comparison of stepper reticle sizes is illustrated in Figure 2-32.

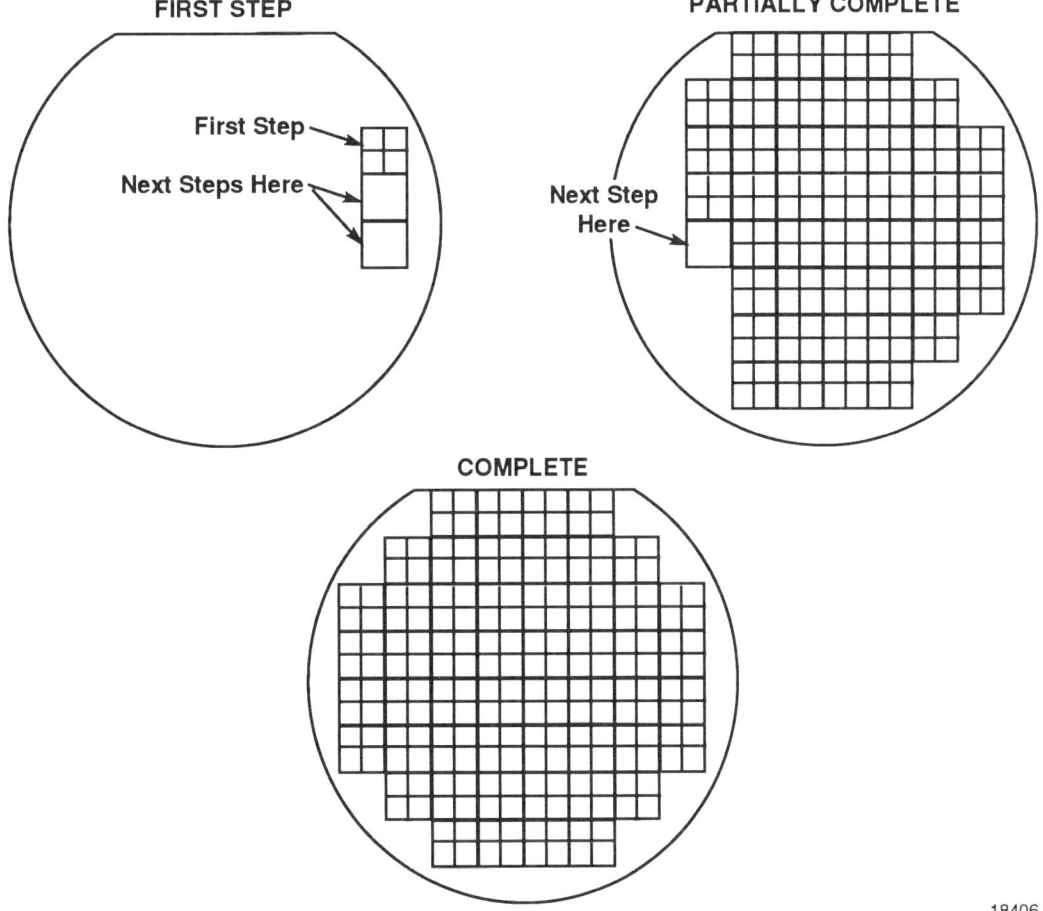

Figure 2-31. Direct Step on Wafer (DSW)

Basic Integrated Circuit Manufacturing

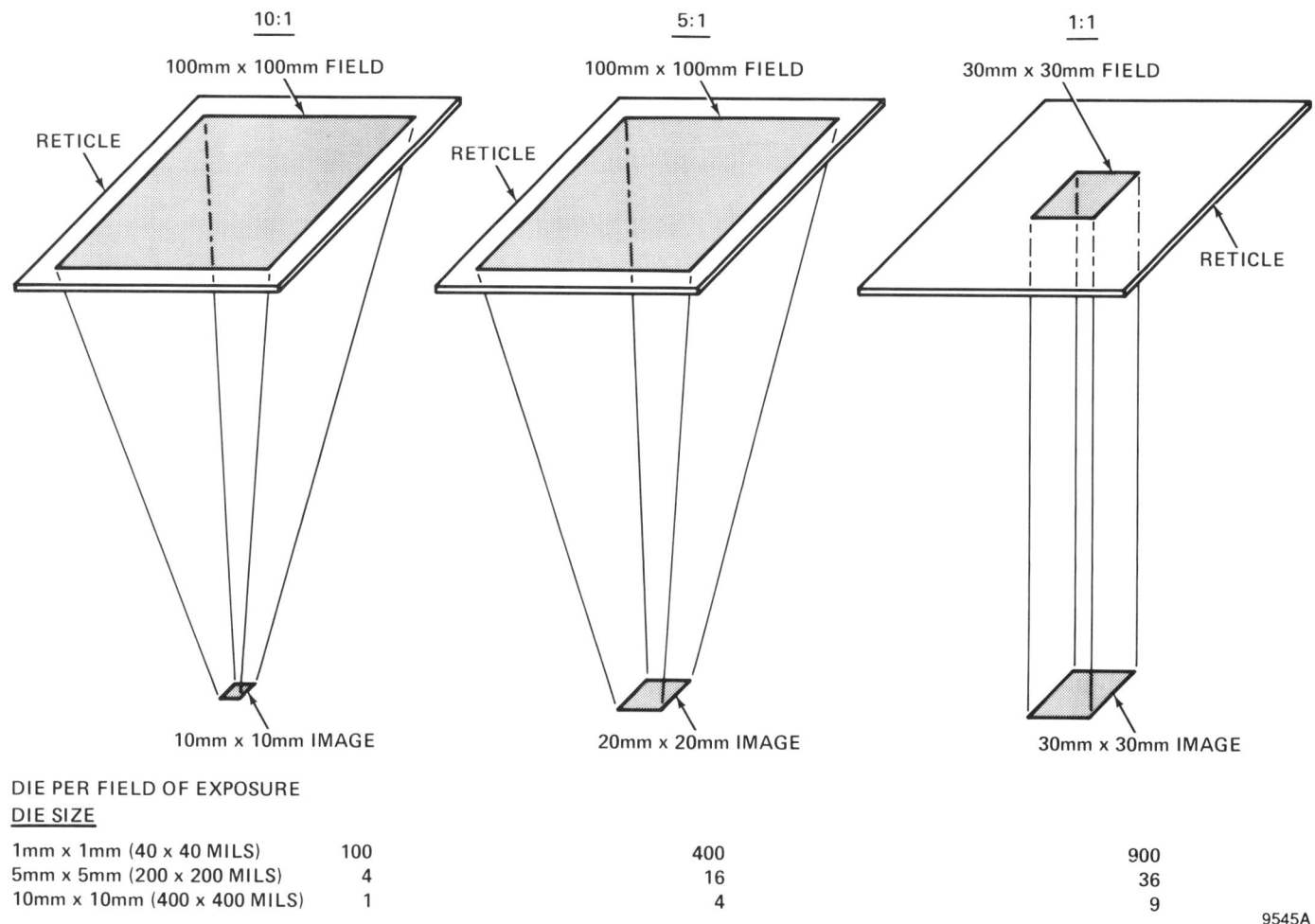

Figure 2-32. Comparison of Stepper Reticle Sizes

- **Pellicles**

The cleanliness of the wafer fabrication area has a significant impact on the defect density of the align/expose process. To improve the defect density of this operation, the reticle is commonly placed in a fixture that has a thin membrane, called a pellicle, stretched over the surface of the fixture. The pellicle then serves as a protective cover over the reticle and de-focuses particles. This prevents the particles from printing. The pellicle is illustrated in Figure 2-33.

The performance of the lens system of DSW equipment continues to improve and the photoresist chemistries continue to be improved to respond to shorter wavelengths of light. The collective result is the capability of manufacturing products, particularly DRAMs, with feature sizes less than 0.5um. As research continues, it may be possible for an optical system to extend to 0.15um features. If this becomes a reality toward the latter part of this decade, then implementation of e-beam and x-ray will be postponed.

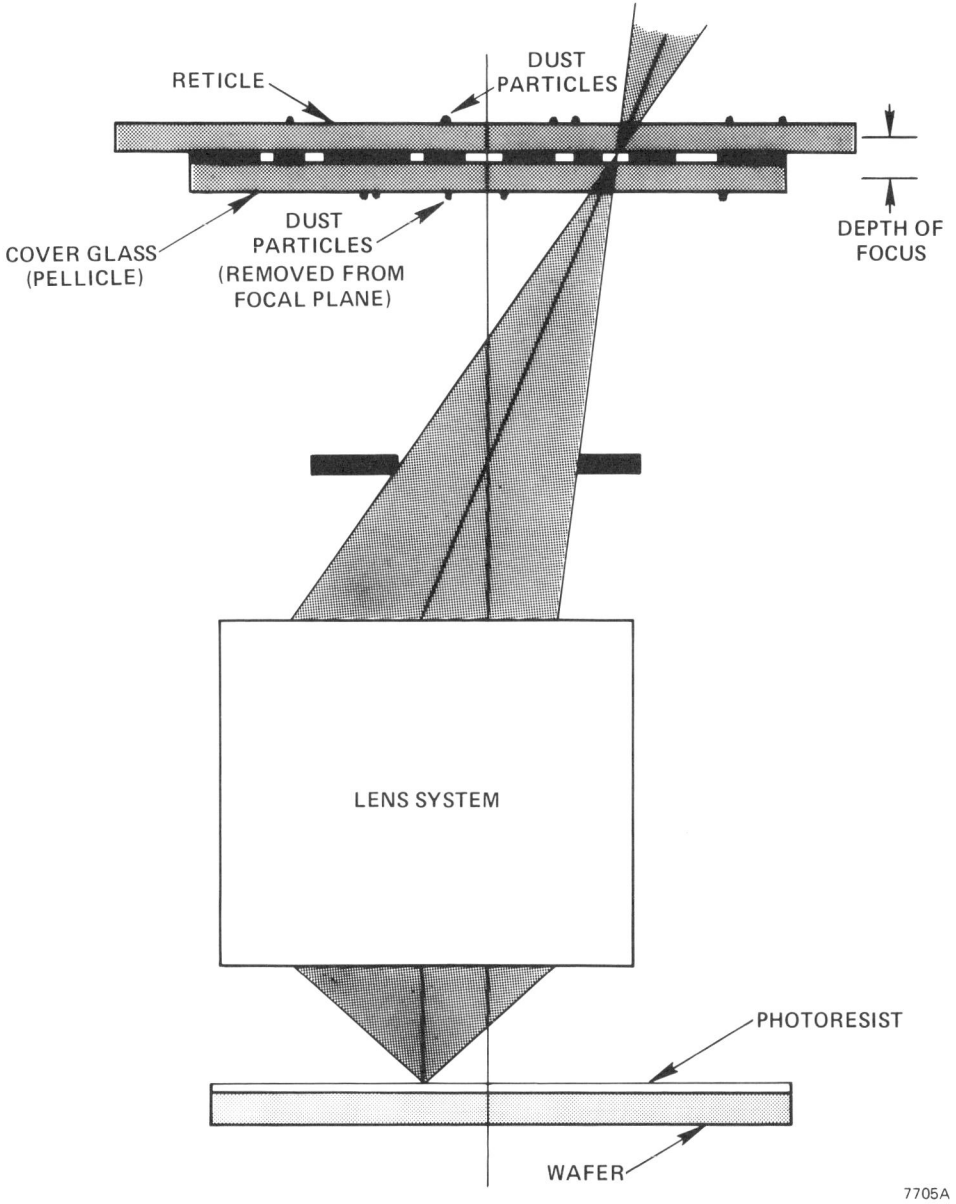

Figure 2-33. Pellicle Protection Mechanism

v. E-beam

As previously shown in Figure 2-20, the electron-beam (e-beam) system has dominated reticle and mask manufacturing. However, direct electron-beam exposure of the resist-coated silicon wafer without a mask or reticle has been very slow to gain production use. The main problems associated with using e-beam wafer exposure systems are related to the poor sensitivity of the photoresists available, the beam current, the electron-electron interaction at the resist-substrate interface, the rate the beam can be moved and the relative high system cost per wafer exposed. The e-beam concept is illustrated in Figure 2-34.

Basic Integrated Circuit Manufacturing

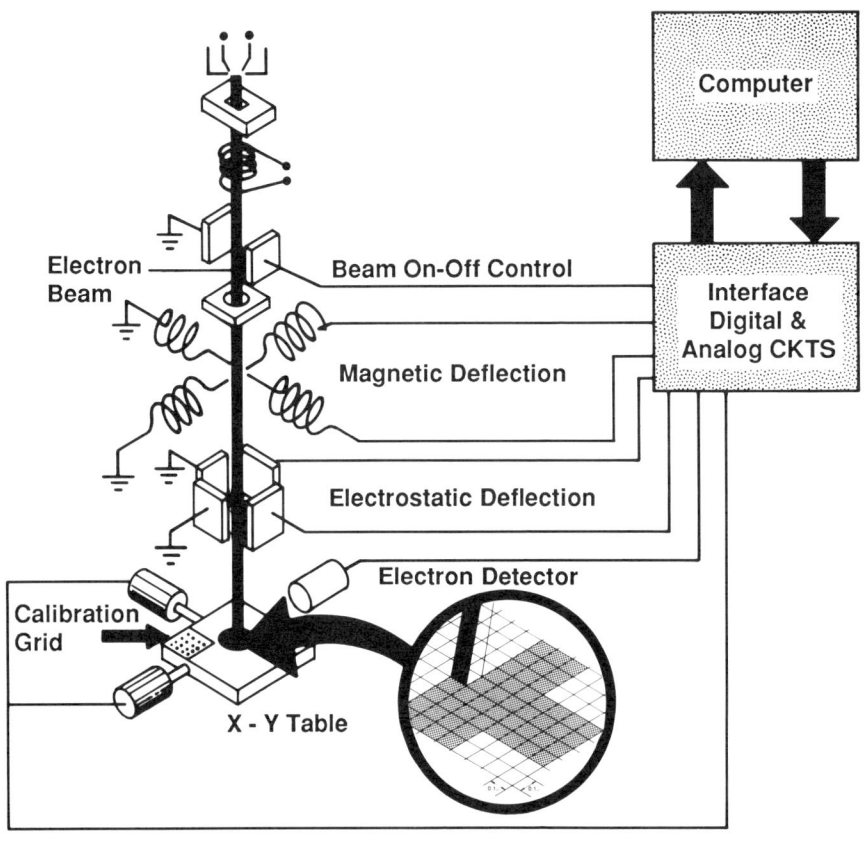

Figure 2-34. E-Beam System

Other disadvantages in the system for e-beam direct-write on the resist-coated silicon wafer are the poor thermal characteristics and the erosion rates in dry etching of the e-beam resist. Both of these parameters are very important in sub-micron lithography.

E-beam systems have been designed using either raster scanning or vector scanning of the beam.

The greatest improvement for the e-beam system will probably come from more compatible photoresist chemistries. This will provide better etch performance. It should be realized that an e-beam system used for reticle and mask masking is not as demanding as writing directly on the wafer. Wafer surfaces are not flat like reticles and masks and therefore create variation in resist thicknesses and scattering effects. Further, the underlying films are considerably thicker making the etching process effects on the resist more severe.

The direct-write e-beam process is essentially a serial process in that each image is formed within a die and then from die to die. Thus, the throughput is very low. This low throughput makes the process very expensive. However, for small volumes, engineering prototypes, or the quick time to market category, the cost may not be the most important consideration. For these needs, the e-beam system is a good choice for align and expose.

vi. X-ray

X-ray energy is another possibility as an align/expose system for imaging on wafers coated with photoresist sensitive to x-rays. X-rays have very short wavelengths thereby making possible improved resolution. Because of the very short wavelength relationship to improved resolution, x-ray technology continues to be actively pursued as an imaging technology for the future.

- **X-ray Sources**

The energy for an x-ray system comes from either high-energy sources like the synchrotron ring or electron impact low-energy sources. The synchrotron ring puts out a wide spectrum of radiation from the infrared region down to the very short wavelengths of x-rays. Beamlines are used to access this form of radiation.

The impact source of x-rays also delivers a broad spectrum of wavelengths by directing an electron beam at a target material. The x-ray radiation given off by the electron-beam impact matches the physical constants of the material and is generally below the intensity of the synchrotron source.

X-ray exposure has been used for full field proximity, 1:1 projection, and step and repeat. As in optical technology, x-ray stepping can take advantage of greater uniformity radiation in a smaller field size and greater intensity of energy. Full field x-ray is optimized for throughput at the expense of overlay registration and pattern resolution. The impact concept is illustrated in Figure 2-35.

- **X-ray Masks and Reticles**

A serious limitation to the x-ray method of image transfer is in the mask or reticle creation. The penetration power of the x-ray energy imposes severe restrictions on the materials used for reticles or masks (Figure 2-35). This causes the tooling cost for the mask or reticle to be three to five times the cost of optical masks or reticles. At this stage of development the slight difference in the spectral output between the synchrotron (0.05 to 0.5nm) and the electron impact (0.8 to 2.2nm) has effect on this tooling cost.

vii. Summary

A comparison of lithography technologies is shown in Figure 2-36. A comparison of throughput, capital cost, feature size and overlay registration is summarized in Figure 2-37.

The image transfer process is constantly undergoing change and refinements. Some of these activities are found in multilayer resist patterning, contrast enhancement techniques, the use of anti-reflective coatings, chemical amplification of the resists and adding chemical dyes to the resists. These techniques and others to follow in the future will allow further improvements to be made. The net result will be smaller feature sizes, higher levels of integration, and more complex integrated circuits.

Basic Integrated Circuit Manufacturing

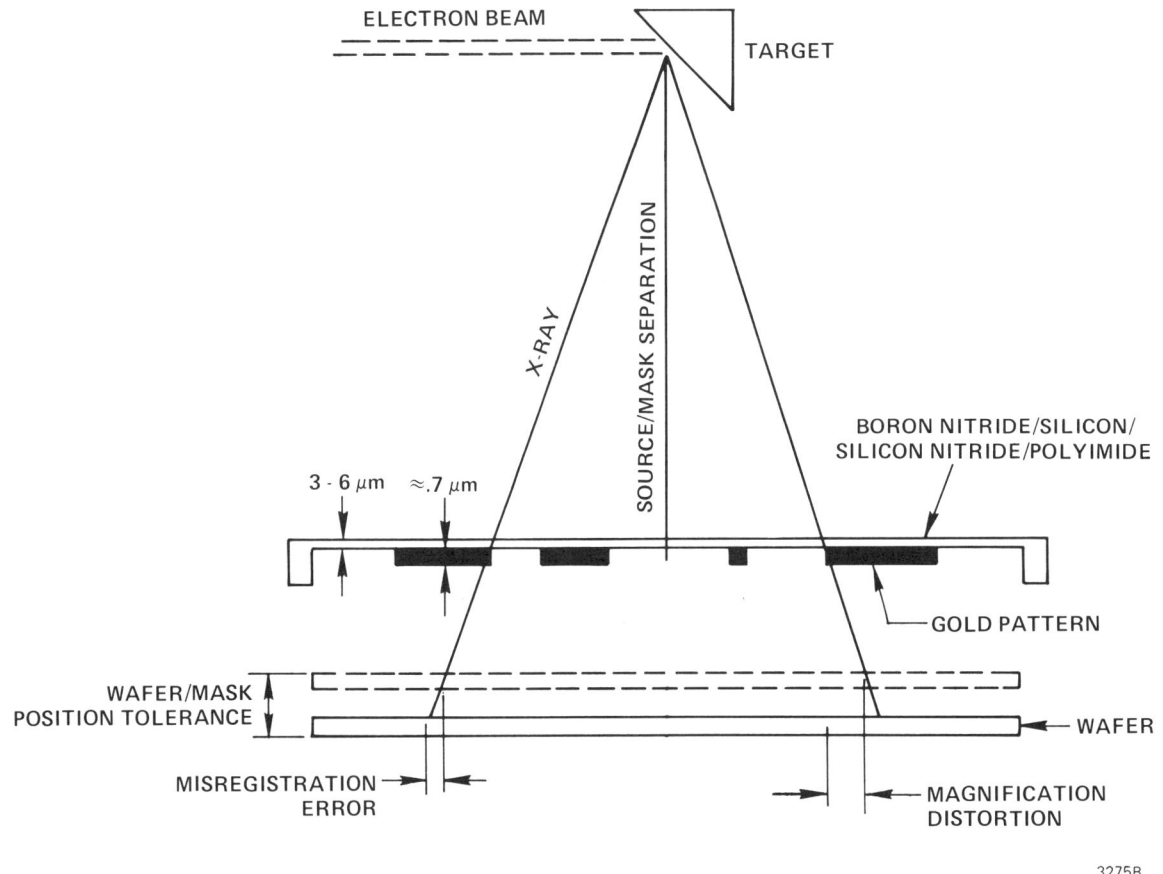

Figure 2-35. X-Ray Lithography

After the align/expose process has been completed, some types of photoresists require a post-exposure bake to further stabilize the film. Otherwise, the wafer is ready for the develop process.

e. **Photoresist Developing**

i. **Positive Resist**

The developer solution reacts chemically with the photoresist to form the image in the photoresist layer. If the resist layer is positive resist, an alkaline developer solution will react only with the exposed resist. This chemical reaction dissolves the exposed resist, leaving the unexposed resist. The chemical reaction of the develop process is stopped by flooding the wafer with D.I. water. The water neutralizes the alkaline solution to stop the chemical reaction.

Basic Integrated Circuit Manufacturing

TECHNOLOGY	ADVANTAGES	DISADVANTAGES
Contact Printing	Inexpensive Low maintenance High throughput	Damages photomask High defects
Proximity Printing	Inexpensive Minimal mask damage High throughput Good depth of focus	Regular adjustment required Limited resolution
Scanning Projection	Good resolution Long mask life Low defects	Expensive Regular adjustment required Temperature sensitive Vibration sensitive Global alignment only
Optical Direct Step on Wafer	Very good resolution Low defects Long mask life Good registration	Very expensive Special mask required Distortion/stepping errors Environmental control required
Electron Beam on Wafer	Excellent resolution No mask required Very low print defects Excellent pattern flexibility	Very expensive Low throughput Special Resist Required Very Complex System
X-Ray	Excellent resolution High throughput	Special mask required Special resist required Global alignment and step & repeat

Figure 2-36. Comparison of Lithography Technologies

MACHINE	MACHINE RATE/HOUR	COST ($)	FEATURE SIZE (μm)	OVERLAY REGISTRATION (μm)
Contact Aligner	60	50K	4.0	2.0
Proximity Autoalign	60	150K	2.5	1.5
P & E Autoalign	60	750K	1.0	0.8
Stepper I 5X	20	1.0M	1.0	0.5
Stepper II 5X	20	1.6M	0.5	0.15
Stepper III 5X	20	2.6M	0.35	0.08
Stepper IV 4X	25	3.0M	0.35	0.08
Step and Scan 4X	30	3.2M	0.35	0.08
X-Ray Stepper	25	5.0M	0.3	0.1
X-ray Stepper II	30	5.2M	0.2	0.08
E-Beam	5	4.0M	0.15	0.08

Figure 2-37. Photolithography Equipment Summary (All Rates Normalized to 150mm Wafer)

Basic Integrated Circuit Manufacturing

ii. Negative Resist

Negative resist requires a solvent solution for developing the image. Xylene is the common developer for negative resist. The exposure process causes the negative resist to polymerize. The polymerization of the resist renders the resist insoluble in the developer. Thus, only the unexposed resist is dissolved and washed away. The chemical reaction of the Xylene requires another solvent to stop the action. N-butyl acetate is commonly used to neutralize and stop the developing action and stabilize the system.

iii. Developing Methods

There are several methods used to develop the photoresist. The particular method chosen is determined by the requirements of feature size and the parameters of the resist system. The more commonly used methods are immersion (batch develop), spray, and a modified spray system known as puddle develop.

- **Batch Develop**

Batch develop is the oldest method. The wafers are placed in carrier that is not affected by the developer chemical. The carrier is immersed in the developer solution for a given period of time. The wafers are then removed from the developer solution and placed into a second solution to stop the develop process and rinse the wafers.

The batch develop process has gone through many changes and refinements over the years but has been gradually replaced by one of the spray methods: spray for negative resist and puddle for positive resist.

- **Spray (Puddle) Develop**

The spray technique has the advantage of processing one wafer at a time with fresh developer solution, improving the uniformity of the process and allowing the handling of the wafers from cassette to cassette. This provides excellent productivity.

The conventional spray system used for negative photoresist must be modified for positive resist. The spray system is an adiabatic process. The change in spray pressure between the nozzle and atmosphere causes too much of a temperature change for consistent develop results with positive resist. To overcome this sensitivity to temperature variation, the puddle technique dispenses the developer solution onto the wafer with a very minimal pressure change. (It is like removing a nozzle from a hose). The solution is allowed to flow over the entire surface of the wafer quickly. The puddle of fluid is allowed to stand on the surface for the required develop time wherein a flow of D.I. water floods the surface to neutralize the developer solution. This process has also been refined by controlled temperature of the developer solution, heated wafer holder, multiple puddles, low-pressure spraying after the puddle step, etc. Consequently, consistent results are achieved with this method.

iv. Plasma Descum

A process used to further improve the surface to be etched after the develop process is called Plasma Descum. The Descum process is done by placing the wafer in a plasma system wherein oxygen is injected into the low-pressure process chamber with an applied R.F. field. The R.F. energy ionizes the oxygen and causes a chemical reaction between the resist and the ionized oxygen. This in effect etches the wafer and removes thin films of resist that may not have been completely removed in the develop process, providing a cleaner surface for the etch process.

v. Spin-Dry Process

Some form of the spin-dry process is used to remove the final rinse solution from the wafer for both the positive and negative develop processes. The wafers are ready for a visual inspection. The result of the develop process for positive resist is illustrated in Figure 2-38.

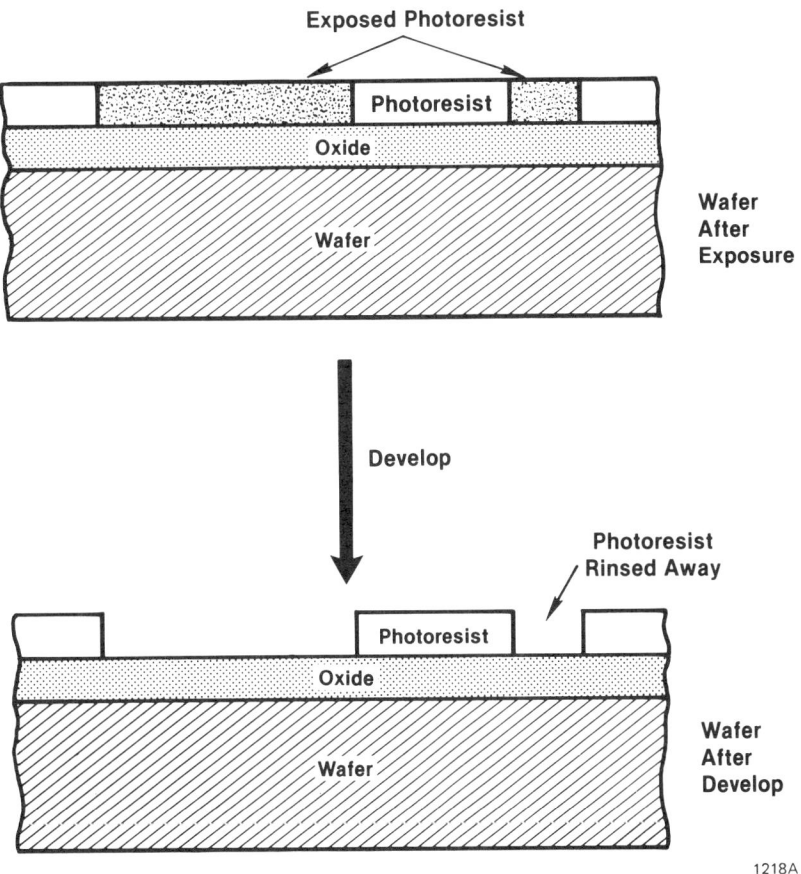

Figure 2-38. Photoresist Develop

vi. Inspection After Develop

Visual inspection of the photoresist-patterned surface is a difficult task. And, as the circuitry becomes denser, the inspection after develop (ADI) will become even more difficult. Therefore, the photoresist process will rely more on Statistical Process Control and less on the visual inspection.

The current trend in technology for monitoring the develop process after developing is to use programmed pattern recognition visual inspection equipment. This type of automation eliminates the eye fatigue problem and provides more consistent results. The automated visual inspection station sample inspects using a statistical plan. This provides immediate feedback for manufacturing control and aids in keeping the rework rate low.

The decision to rework a wafer (or wafers) after the develop inspection has a significant impact on the cycletime of the overall process. Good manufacturing practices dictate keeping reworks low at this inspection. The long-term goal is to be able to eliminate this inspection through continued improvement by using Statistical Process Control.

vii. Post-Develop (Hard) Bake

After the inspection is completed, the wafers are given a "hard bake" at a temperature of 130°C to 160°C to dehydrate the photoresist prior to etch or ion implant processes. This bake stabilizes the resist characteristics and makes the resist less sensitive to these hostile process environments.

f. Etch

The etching process removes the material not protected by the hardened photoresist. The result of the etching process was illustrated in Figure 2-15 for positive resist and is shown in Figure 2-39 for negative resist.

These figures are idealistic representations of the patterning process. In actual manufacturing, there are varying degrees of acceptable results. Figure 2-40 illustrates several examples.

The etching process is divided into two basic methods: isotropic and anisotropic. These terms describe the geometrical shape the film will have after the etching process.

i. Isotropic/Anisotropic

An isotropic etch process will etch laterally while etching the material vertically. The worst case is when the lateral rate is equal to the vertical etch rate. The anisotropic etch will etch only in the vertical direction. These concepts are illustrated in Figure 2-41.

Basic Integrated Circuit Manufacturing

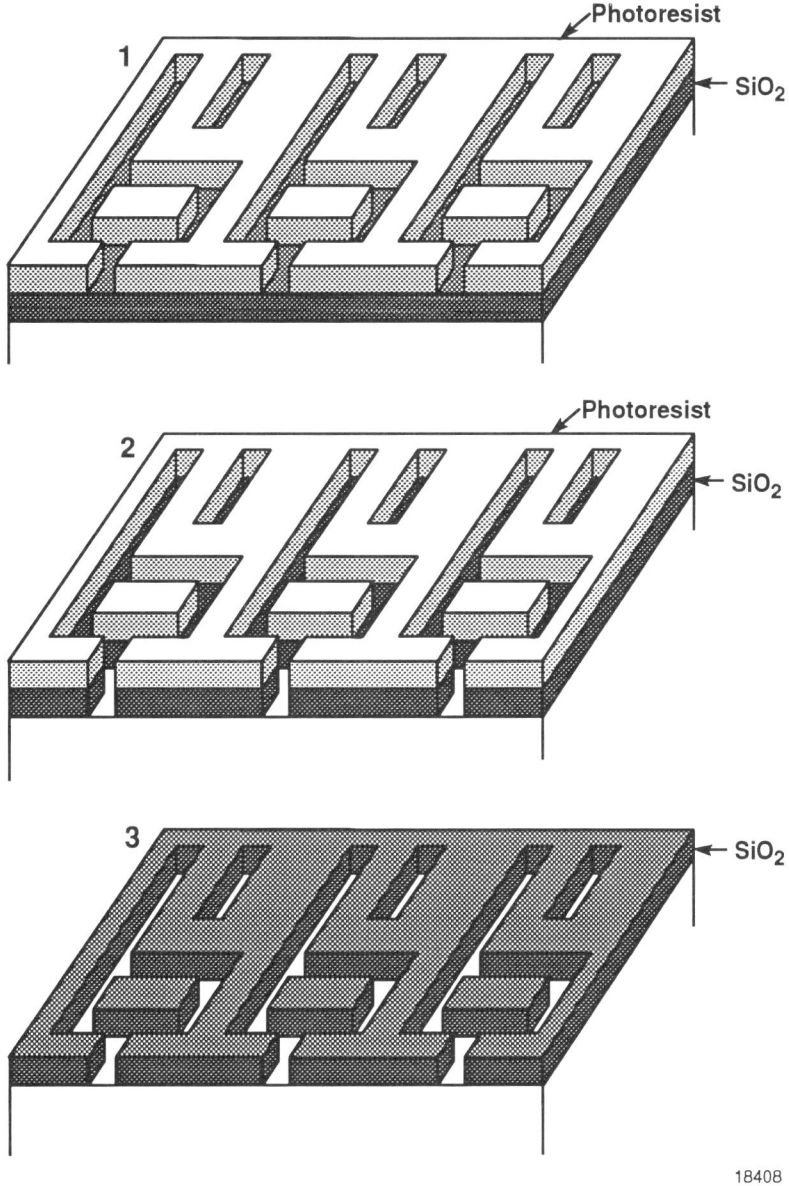

Figure 2-39. Photolithography Using Negative Photoresist

The etching process is essentially a subtractive process to remove the material not protected by the photoresist. There are several methods used to remove any given material. The more common methods are: wet chemistry, plasma dry chemistry, reactive ion etching dry chemistry, and ion milling.

Basic Integrated Circuit Manufacturing

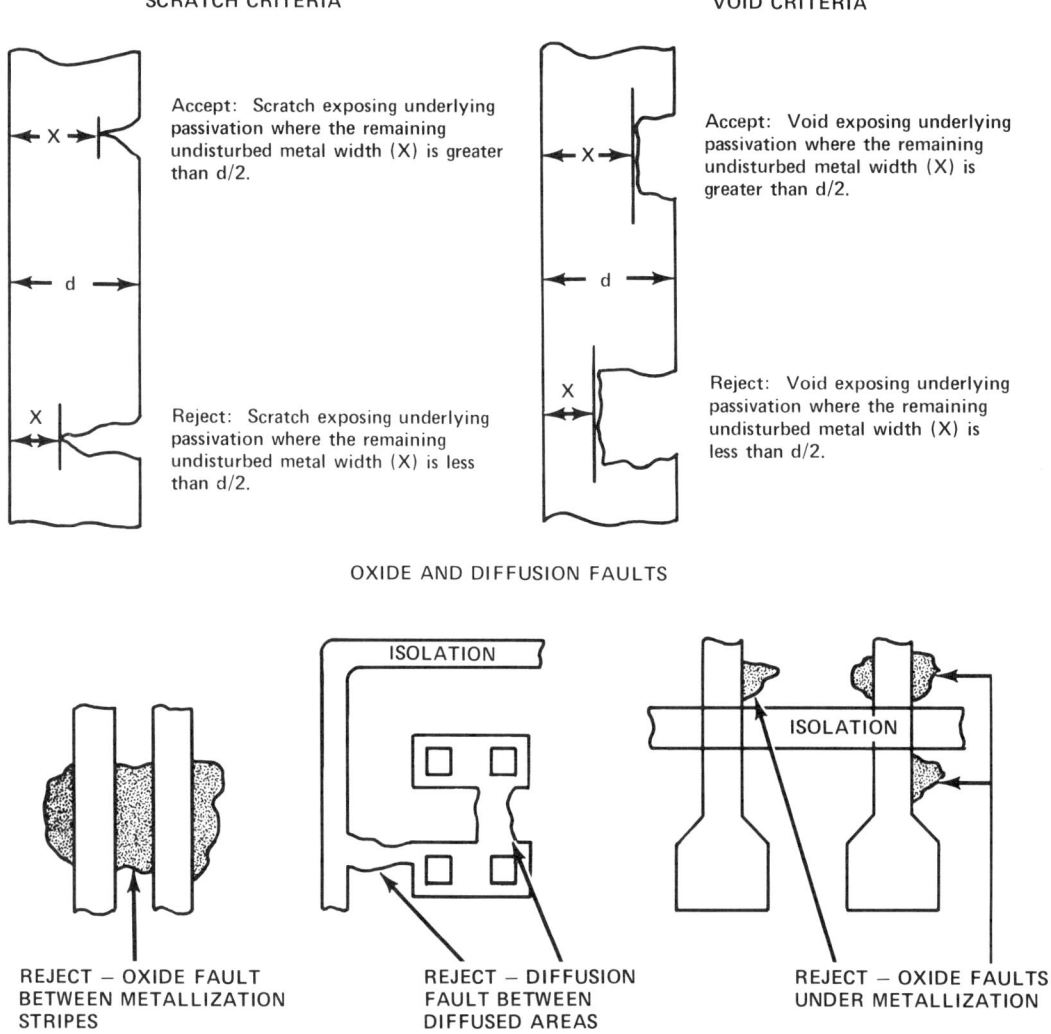

Figure 2-40. Major Visual Defect Examples

Wet chemical etching is the oldest etch method used. A chemical solution that rapidly etches the layer to be removed while having minimal reaction with the photoresist is chosen. This interaction of the etchant to photoresist vs. the layer to be etched is called selectivity. Therefore, the higher the selectivity of the etchant or etch system, the better.

ii. Dry vs. Wet Etch

Feature size has a significant impact on the choice of wet vs. dry etching technique and equipment (Figure 2-42). Generally, feature sizes greater than three microns can be resolved with wet etching. Feature sizes less than three microns usually require dry etching.

Basic Integrated Circuit Manufacturing

ETCH:
The process of removing material — such as oxide or other thin films — by chemical, electrolytic or plasma–ion bombardment.

ISOTROPIC ETCH:
An etch that proceeds laterally and vertically at the same time, sometimes at equal rates. Wet chemical etches are usually isotropic in nature.

ANISOTROPIC ETCH:
An etch that has very little lateral activity. Most of the etching occurs in the vertical direction. Dry etch systems are primarily anisotropic.

Figure 2-41. Vocabulary – Etch Process Section

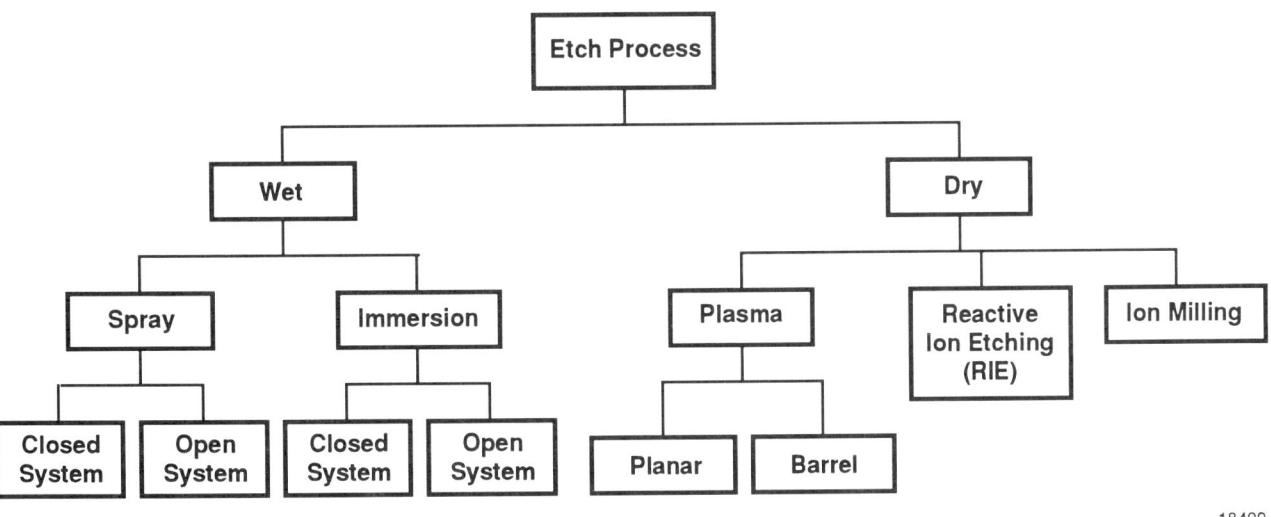

Figure 2-42. Etch Methods

iii. Wet Etch

Wet chemical etching of oxides is done with Buffered Oxide Etch (BOE). This etchant is composed of ammonia fluoride (NH_4F) and hydrofluoric acid (HF). The concentration of the HF is varied to affect the etch rate of the solution.

Silicon-nitride films are etched with hot phosphoric acid (H_3PO_4). Generally, the temperature is in the 150°C - 180°C range. Because photoresist does not stand up well at this temperature, thin layers of silicon dioxide are used on top of the silicon nitride as the etch mask.

Aluminum films are etched with a mixture of phosphoric acid, acetic acid (HAC), and nitric acid (HNO_3) diluted with D.I. water. The etch is used in the temperature range of 20°C - 80°C.

Monocrystalline and polycrystalline silicon are etched with a mixture of HF, HNO_3, and HAC. The temperature of the solution is maintained in the 20 - 30°C range.

Wet chemical etching can be done by either immersion in a given chemical solution or by spraying the chemical onto the surface. After the etching process is completed, the wafers are rinsed with D.I. water and dried in a spin-dryer.

Unfortunately all of the wet chemical etches leave the edge of the film removed with some degree of slope.

iv. Dry Etch

The choice among the use of the various dry etch methods is feature-size driven. The more critical the tolerance, the more likely RIE will be chosen over plasma methods and hardware.

Dry etch processing is performed under some degree of vacuum (Figure 2-43). Dry etch systems require pumping systems for low-pressure operation and gas flow control devices for the various etchants along with an R.F. power supply and control electronics. These etch systems are more complex and require more maintenance than the simpler wet etch stations.

Barrel-type reaction chambers were initially used for plasma etching. This design was acceptable for removing photoresist but had poor uniformity characteristics for dielectric etching. The plasma barrel-type etcher is shown in Figure 2-44.

The next generation design was the planar parallel-plate plasma reactor. This improved uniformity of etching dielectric layers and led to the lower operating pressure reactive ion etcher (RIE). This is shown in Figure 2-45.

Ion milling systems are still being perfected. The selectivity of the system is poor and the capital cost is high.

g. Photoresist Removal

At the completion of the etching process, the photoresist is removed. The pattern has now transferred from the mask or reticle into the resist and from the resist into the layer below the resist (Refer back to Figures 2-15 and 2-39). This basic process is repeated for all of the number of layers a particular design requires.

Basic Integrated Circuit Manufacturing

< 100 MILLITORR

PHYSICAL SPUTTERING
(and Ion Beam Milling)

- PHYSICAL MOMENTUM TRANSFER
- DIRECTIONAL ETCH (ANISOTROPIC) POSSIBLE
- POOR SELECTIVITY
- RADIATION DAMAGE POSSIBLE

HIGHER EXCITATION ENERGY

100 MILLITORR RANGE

RIE (Halocarbon Gas)

- PHYSICAL (ION) AND CHEMICAL
- DIRECTIONAL
- MORE SELECTIVE THAN SPUTTERING

HIGHER PRESSURE

PLASMA ETCHING

- CHEMICAL, THUS FASTER BY 10–1000X
- ISOTROPIC
- MORE SELECTIVE
- LESS PRONE TO RADIATION DAMAGE

Courtesy of MRC

Figure 2-43. The Dry Etching Spectrum

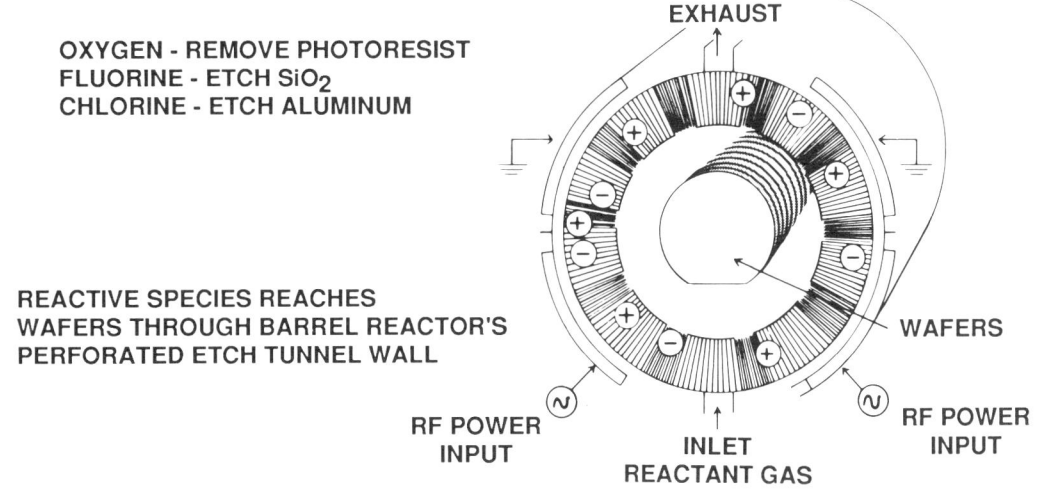

OXYGEN - REMOVE PHOTORESIST
FLUORINE - ETCH SiO_2
CHLORINE - ETCH ALUMINUM

REACTIVE SPECIES REACHES WAFERS THROUGH BARREL REACTOR'S PERFORATED ETCH TUNNEL WALL

Figure 2-44. Diagram of Barrel-Type Plasma Etcher

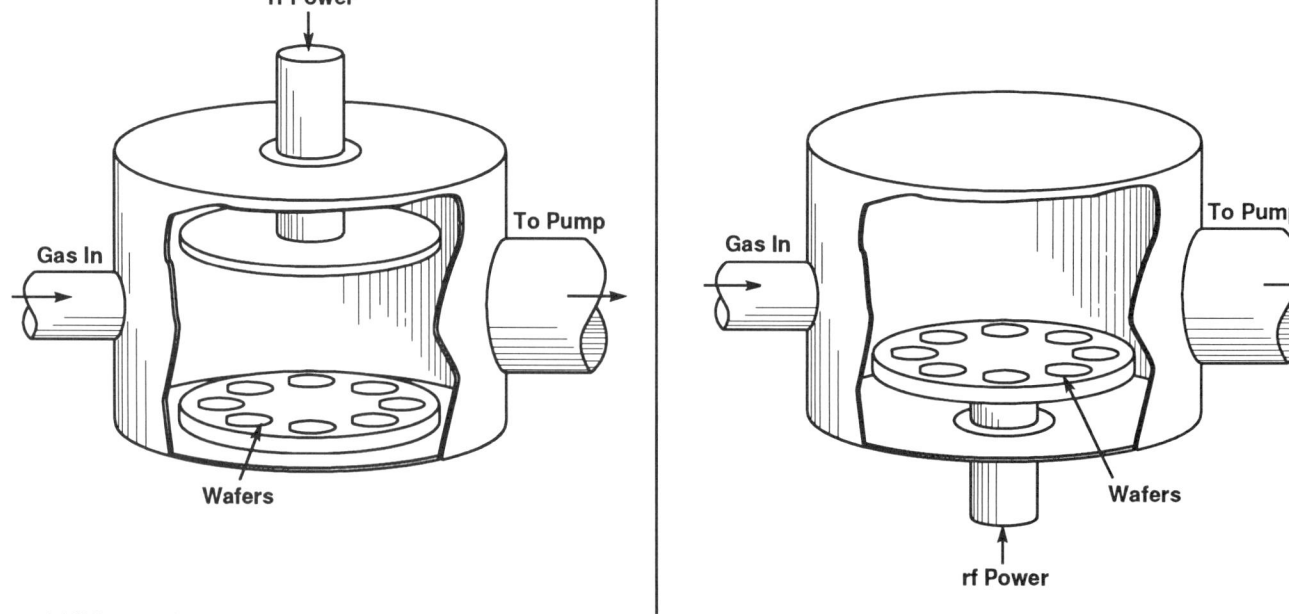

(a) When wafers are placed on the grounded electrode, the system is configured in the plasma etch mode.

(b) When wafers are placed on the powered electrode, the system is operated in the reactive ion etch, or RIE, mode.

Source: Silicon Processing for the VLSI Era

Figure 2-45. Planar Type Reactor

The actual removal of the photoresist can be done in several different ways.

1. Wet chemistries
2. Dry chemistries
3. Heat
4. Light energy and gases.

Each removal technique has advantages and limitations.

Photoresist used for layers prior to metallization can be removed with wet chemistries such as sulfuric acid (H_2SO_4) and an oxidizer like peroxide or ammonium persulfate, sulfuric acid with chromic acid, organic acids, and solvents.

Dry chemistry removal usually uses oxygen in the presence of R.F. energy. The reaction generates gaseous oxides of carbon and water. These species are pumped away by the vacuum pump. Unfortunately, there are a few metallic ions present in the resist system that do not get removed with this cleaning method. However, this can be overcome with a wet chemical clean. In the case where a metal conducting layer is the layer under the photoresist, organic acids can be used as the follow-up clean.

Heating the photoresist as a cleaning process is not used because of the residue remaining. A technique that is gaining acceptance is to use ultraviolet energy in the presence of ozone (O_3).

After the photoresist has been stripped (removed), the wafer is given some type of visual inspection. Like other visual inspections, pattern recognition inspection equipment and sampling techniques are used.

E. JUNCTION FORMATION

Recalling from the materials section, dopant atoms are added to the silicon lattice when the silicon ingot is grown to provide the electrical characteristics of the wafer (N-type or P-type). During the process of producing ICs on the wafer, atoms of the same polarity as the wafer or of the opposite polarity must be introduced into the wafer in selected regions. This alteration of the dopant levels is done by solid-state diffusion or ion implantation.

1. Diffusion

Diffusion is the term used to describe the movement of atoms, molecules, or particles from a location of high concentration to a new location of lower concentration. An example of the process of diffusion can be visualized by watching a small drop of red coloring being dropped into a glass of clear water.

Initially, as the drop strikes the clear water, the red color is highly concentrated. After a short period of time the deep red color begins to lessen and the color starts to spread out over a larger volume. As more time elapses, the color continues to spread out and the red color starts to change to pale red and on into a pink. In a given time the pink will almost disappear in the now nearly clear water.

The process just described represents diffusion as a function of time and temperature. The rate an atom, molecule or compound diffuses from a high concentration to a lesser concentration is related to temperature and time. The parameter that ties the diffused rate to temperature for a given time is known as the diffusion coefficient or diffusivity.

The dopant atom must displace a silicon atom in the crystal structure of the silicon to become electrically active. The process of diffusion is used in semiconductor processing to introduce a controlled amount of a chosen dopant into selected regions of a semiconductor crystal. The diffusion process used to accomplish this substitution is divided into two distinct steps.

1. Predeposition
2. Redistribution or drive-in

a. Predeposition

The appropriate diffusion hardware is chosen to allow the predeposition process to introduce a carefully controlled amount of the desired dopant into the semiconductor crystal. The variables in the process and selection of the dopant species are: the rate the species diffuses into the silicon at a given temperature, the amount of dopant the silicon will accept into the crystal structure at a given temperature, and the amount of silicon dioxide required to mask the dopant at the given process temperature. The required transistor parameters determine how many dopant atoms should be in the silicon and how far the dopant concentration should be below the surface. This selective doping process is shown in Figure 2-46. The illustration is also indicating that the dopant source is available during the entire predeposition cycle.

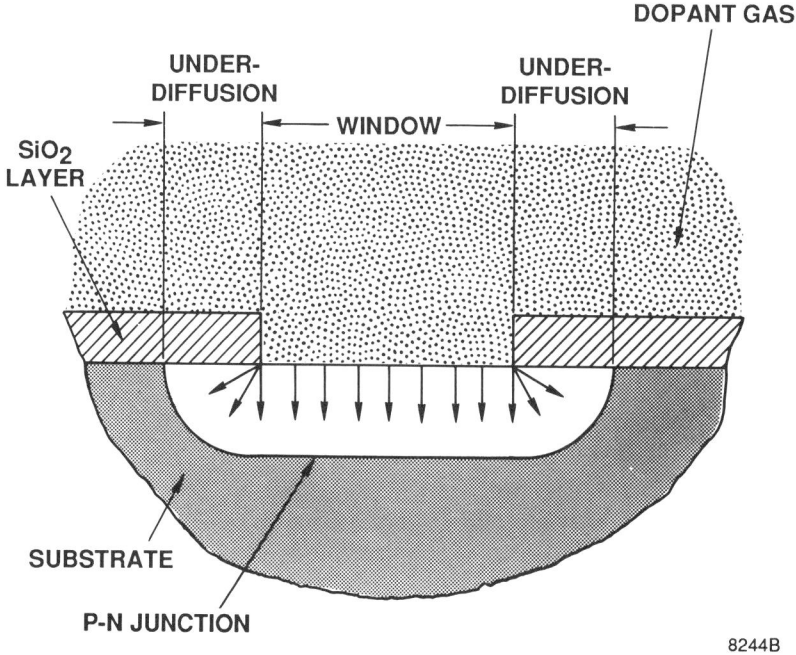

Figure 2-46. Selective Doping Using SiO$_2$ as a Mask

The dopant species can be brought to the wafer surface several different ways. Liquid chemical sources can be introduced into the process tube as controlled chemical vapors or applied directly to the surface prior to loading the wafers onto the quartz diffusion boat. Gas sources can be precisely metered into the process tube. Solid diffusion sources made from high-temperature materials can be loaded onto the quartz boat with the patterned side of the wafer facing the source material. Each of these methods has its own set of advantages and limitations.

Basic Integrated Circuit Manufacturing

The predeposition process will form a thin layer of doped silicon dioxide on the wafer surface during the predeposition cycle. This oxide has incorporated into its structure the dopant. This doped oxide layer is generally removed prior to the redistribution process. The removal is done by wet chemical methods or a sequence of a low-temperature thermal oxidation for a brief time followed by the wet chemical method.

If the source is made available for an extended period of time, more and more dopant atoms will be added to the silicon lattice. As this diffusion process is allowed to continue longer and longer, the concentration gradient goes in deeper and deeper. If left to continue, eventually the silicon wafer would be completely doped uniformly throughout the thickness of the wafer. This is illustrated in Figure 2-47.

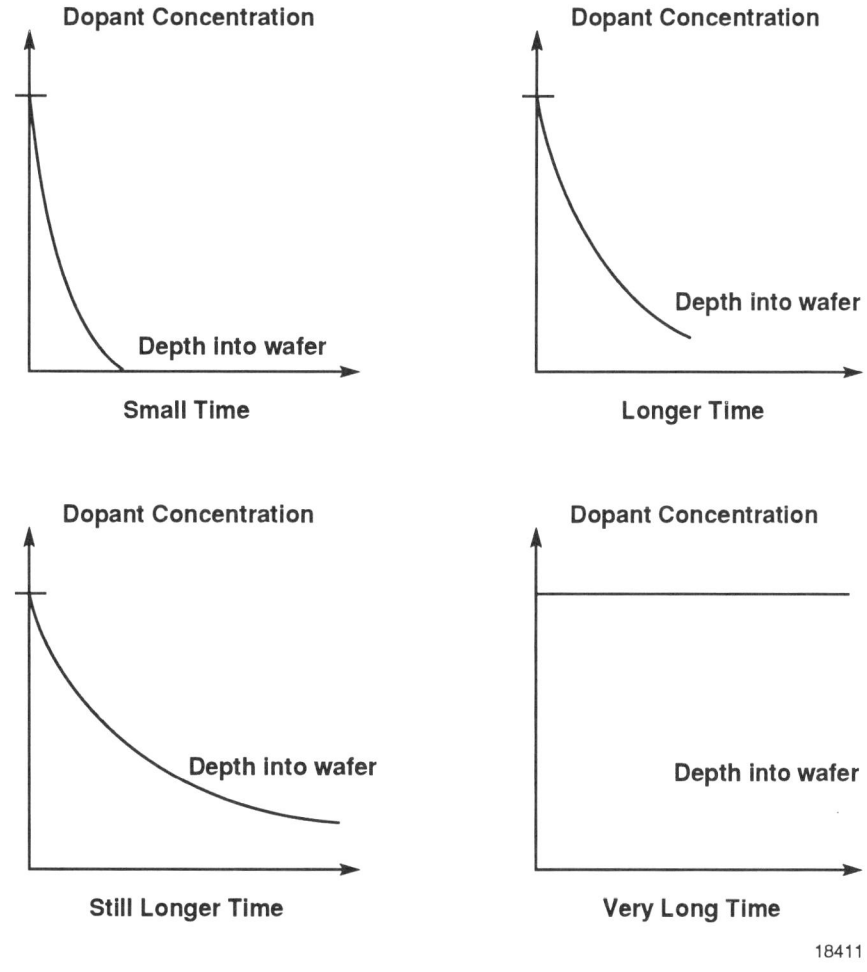

Figure 2-47. Profile of Dopant Present in a Wafer as a Function of Time

To control the device parameters, the predeposition process embeds a precise quantity of dopant atoms into the silicon using a dedicated high-temperature furnace. At the completion of the predeposition process, the wafers are removed from this furnace and moved to a different furnace for the second step of the process.

b. Drive-In

The redistribution or drive-in process is a diffusion step in which no additional dopant is introduced into the wafer. This process step is done in a controlled atmosphere of nitrogen, nitrogen and oxygen, steam, or combinations of these. At some part of the redistribution cycle, an oxidizing atmosphere is used to regrow a protective oxide layer over the diffused regions. The process variables of time, temperature, and ambient gases are controlled. These variables control the final junction depth, the thickness of the oxide in the diffused region, and the exact dopant profile in the wafer.

The predeposition and redistribution process flow charts are shown in Figures 2-48 and 2-49. In wafer fabrication, both processes are performed in furnace equipment similar to that shown in Figure 2-50. Wafers that are ready for predeposition are cleaned to remove any contamination from storage or prior operations.

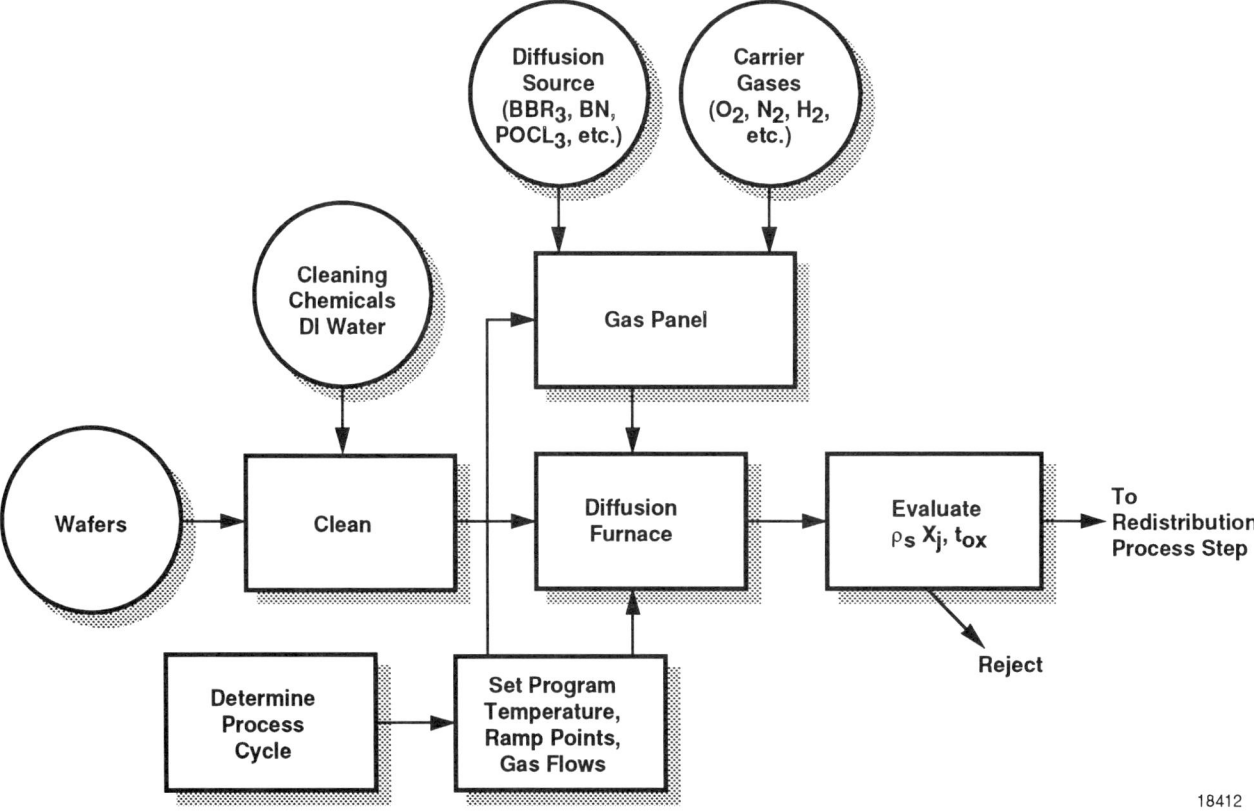

Figure 2-48. Diffusion Process Flow Chart for Predeposition

Basic Integrated Circuit Manufacturing

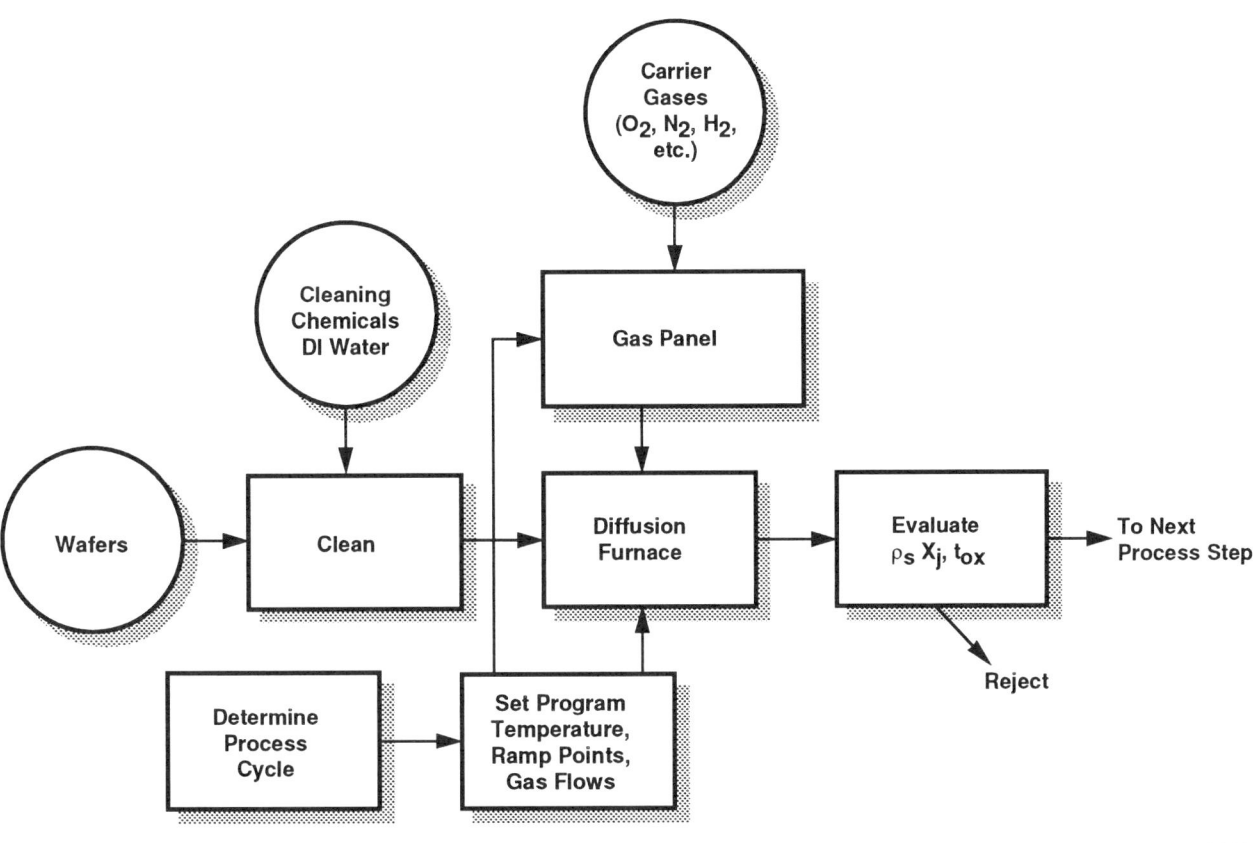

Figure 2-49. Diffusion Process Flow Chart for Redistribution

c. Process Control Measurements

Process control measurements are performed after both predeposition and redistribution (drive-in). Because of the selective nature of these process steps, "test wafers" are used in both the pre-dep and redistribution process steps for these process control measurements. Typically, the sheet resistances of the test wafers are recorded after the predep step and the test wafers are placed on the redistribution boat. After the redistribution process is completed, the new sheet resistances, oxide thicknesses, and junction depths are measured and recorded. These measurements are assumed to be representative of the actual product wafers. For correlation purposes, the product wafers have test patterns that are measured at the completion of the process. These readings are then compared to the test wafers' readings. The correlation is good.

Basic Integrated Circuit Manufacturing

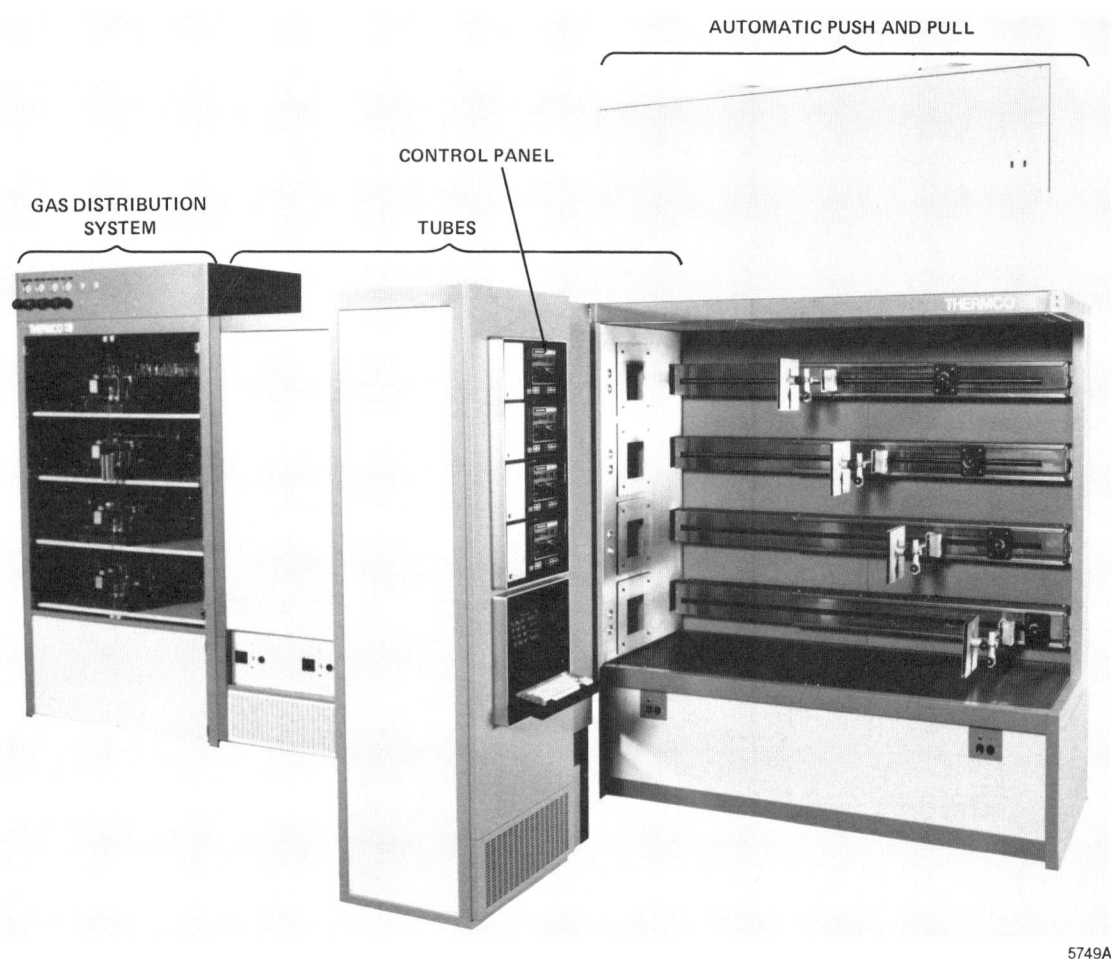

Figure 2-50. Oxidation/Diffusion Systems

i. Resistivity

The resistivity measurement made on the test wafer is illustrated in Figure 2-51. This type of resistivity measurement is often measured in terms of ohms per square, R_s or $R_\square$. For the diffusion process, this means that the test wafer used is doped with the opposite type carriers, i.e., a P-type wafer is used for N-type diffusions and an N-type wafer is used for P-type diffusions. The ohms per square concept is further illustrated in Figure 2-52. In this figure, the measuring current, I, is being forced through thickness, Xj, the diffused region. As long as the lateral dimensions are large and the thickness, Xj, is small compared to the probe needle spacing, s, the sheet resistance, R_s, can be determined. In Figure 2-52, L_1 equals L_2, and R_s is the ratio of the measured voltage, V, to the forcing current, I, multiplied by the constant, 4.532. In practice, the current, I, is decade multiples of 4.53. This makes the measured voltage, V, a direct reading of resistance in ohms per square.

This reading has sensitivity to light. Therefore, the readings are generally taken in a dark enclosure.

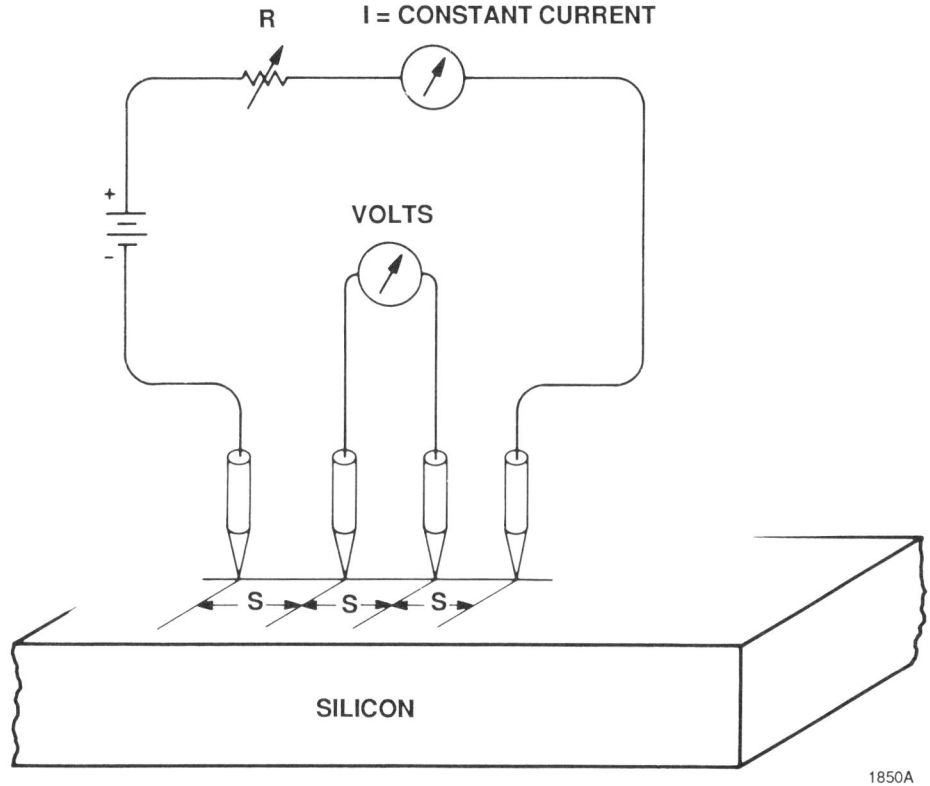

Figure 2-51. Resistivity Measurement (ρ)

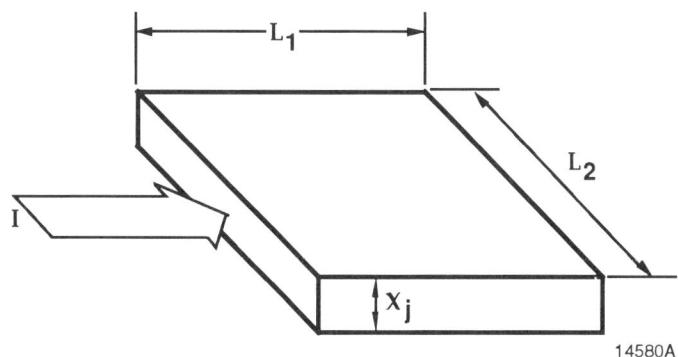

Figure 2-52. Illustration for Ohms per Square

ii. Junction Depth

The junction depth measurement, **Xj**, is made on a test wafer by lapping the wafer at a very shallow angle. The angle lapping concept is illustrated in Figure 2-53. After lapping, the wafer is etched in a chemical that stains the P region darker than the N region. This difference in color will reflect monochromatic light such that the interference fringes can be counted. Knowing the wavelength of the monochromatic light and the number of fringes allows the junction depth to be determined. Figure 2-54 shows the results of junction staining and monochromatic light on a bipolar transistor. Figure 2-55 illustrates an MOS transistor with the source and drain junctions delineated.

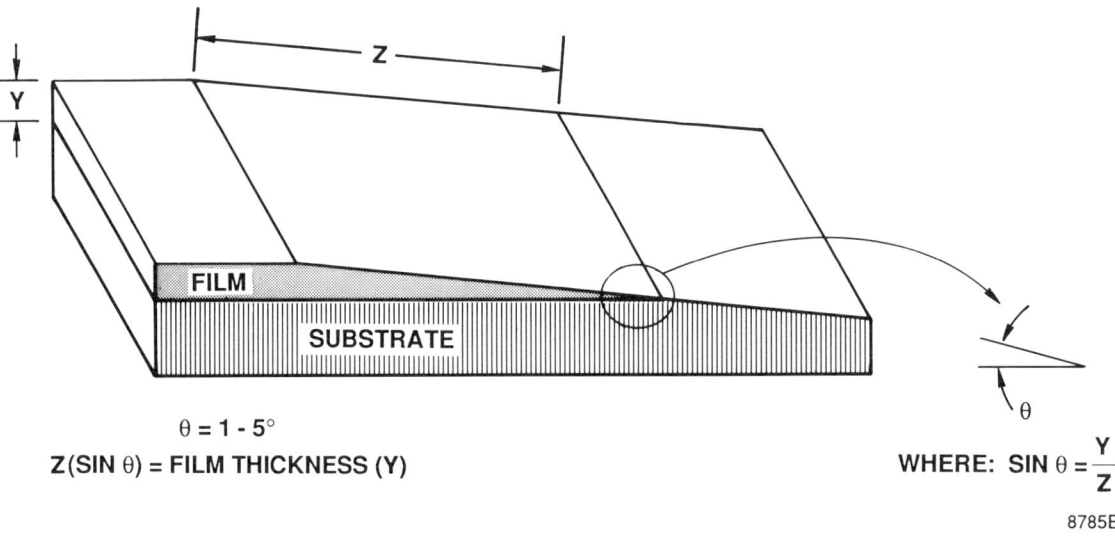

$\theta = 1 - 5°$
$Z(\sin \theta) = $ FILM THICKNESS (Y)

WHERE: $\sin \theta = \dfrac{Y}{Z}$

Figure 2-53. Angle Lapping Measurement Method

2. Ion Implantation

An alternative method for introducing dopant atoms into silicon emerged in the late 1960's. This new development is called ion (an ion is a charged atom) implantation. Essentially, ion implantation is the process of shooting ions of the desired dopant species into the wafer through openings in the oxide or hardened photoresist.

In contrast to diffusion, ion implantation is a low-temperature technique. It provides a flexibility not available with diffusion. The ion implant process takes ions of a desired dopant, accelerates them using an electric field, and scans them across the wafer surface to obtain a uniform predeposition. Scanning is the process of the beam moving back and forth across the surface of the wafer. This predep concentration is similar to the predeposition described in the diffusion section, except the dopant does not displace silicon atoms in the silicon lattice after the implant is completed. More on this in a later paragraph.

Basic Integrated Circuit Manufacturing

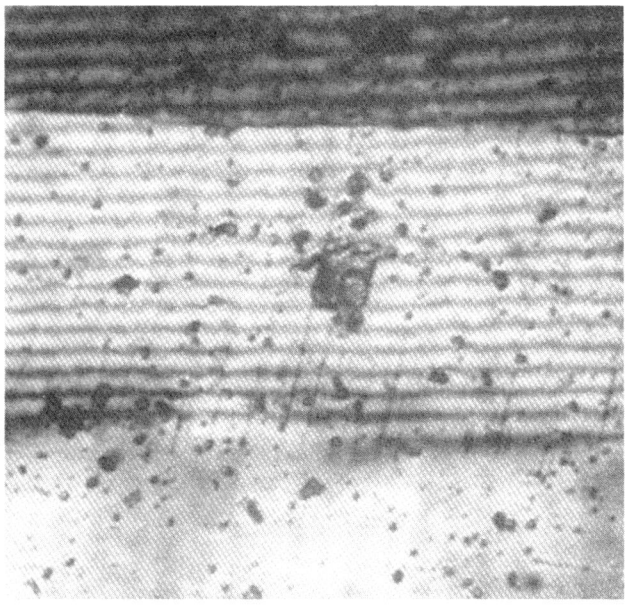

A. Stained single-junction structure revealed by interference fringes.

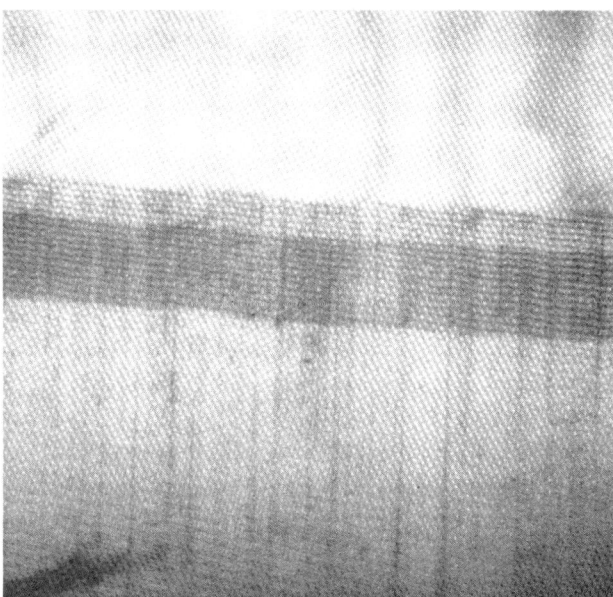

B. Stained double-junction structure revealed by interference fringes.

1851

Figure 2-54. Monochromatic Light Examinations

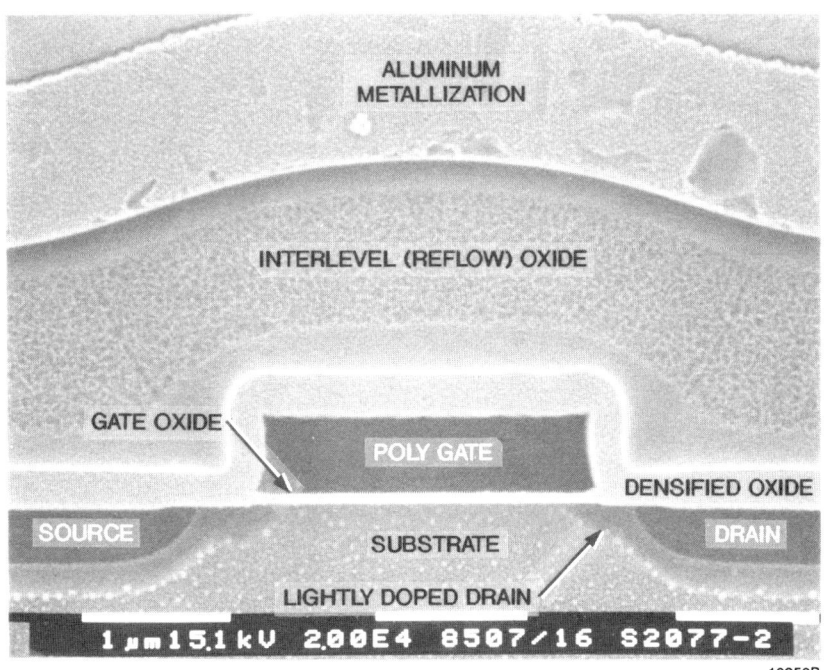

Figure 2-55. Silicon Gate MOS Cross-Section

Basic Integrated Circuit Manufacturing

The first requirement of an ion implantation system is the ability to generate ions of the desired species. A gaseous source containing the dopant atom is fed into a vacuum chamber and ionized by coming into contact with a hot filament. The correct ions are separated from any other ion generated by bending them a set angle using an electromagnetic field. This separation by atomic mass allows only the desired ions to enter the acceleration tube. The selected ions are accelerated using an electric field and strike the target wafer at some predetermined high-energy level. The ion beam passes through an electro-optical beam shaping region of the accelerator tube to control the diameter of the beam and causes the beam to be scanned. It is a scanned beam that strikes the wafer surface and penetrates inward to some depth. The basic concepts are summarized in Figure 2-56. A diagram of an ion implanter is shown in Figure 2-57.

1. Generation of ions in some kind of source region.

2. Separation of ions by mass to prevent undesired species from reaching the target.

3. Acceleration of the selected ions to high energies.

4. Bombardment of the sample (located in a target chamber) by the high-energy ions.

Figure 2-56. Basic Concepts of Ion Implant

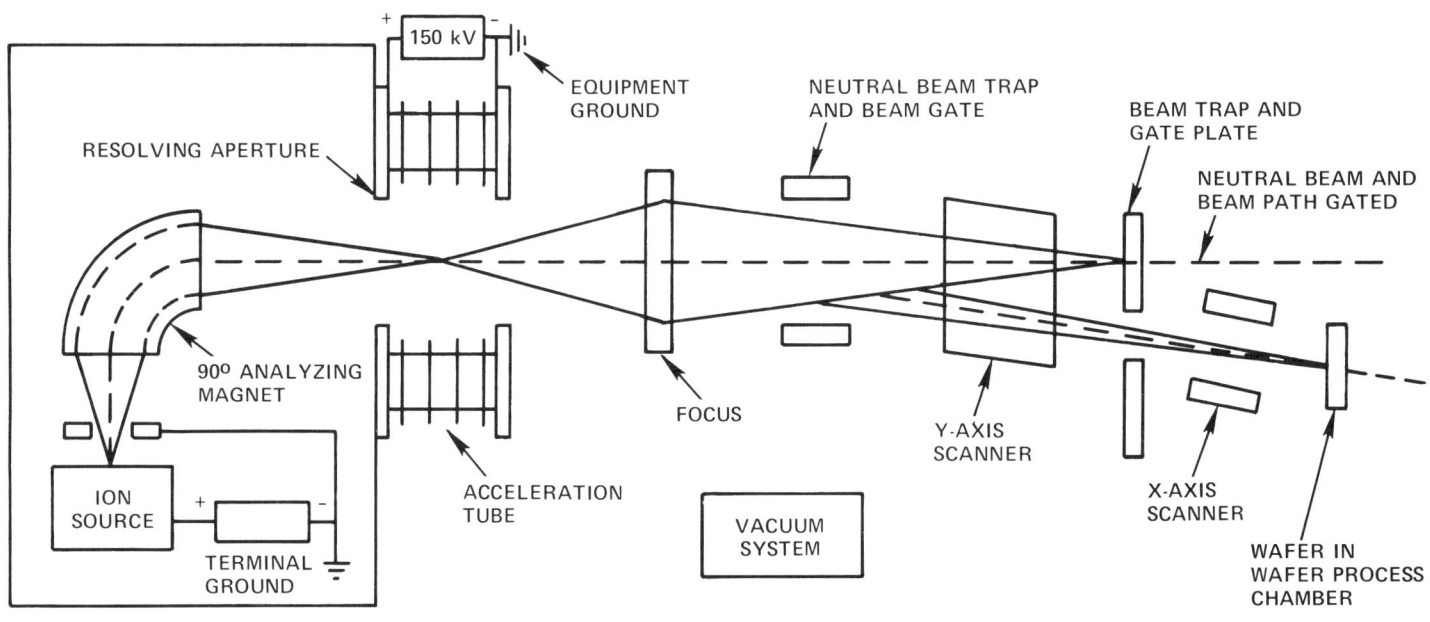

Figure 2-57. Configuration of a Typical 200KeV Ion Implanter

Basic Integrated Circuit Manufacturing

The ion implanter is a large, complex, and expensive piece of fab equipment. It contains several vacuum systems, a gas distribution system, high-voltage power supplies, complex cooling systems, and computer-controlled electronic systems. For all of this, the fab gets a control of the dopant process that can not be obtained any other way. The ion implanter has played such an important role in semiconductor fabrication that MOS technology would probably not dominate the IC industry without it.

Once the doping species has been selected, the two variables that can be controlled are the dose (the number of ions that strike the wafer surface per unit area) and the acceleration energy (which determines the depth of the implant into the wafer). The dose is controlled by counting the ions as they pass a detector, and the acceleration energy is controlled by changing the voltage on the accelerator tube. The ability to control both dose and energy provides a powerful and unique tool for semiconductor manufacturing.

The results of the implant process is a silicon lattice that has been disrupted and damaged. The path of an implanted ion is shown in Figure 2-58. A heat treatment (anneal) must be performed to reorder the silicon lattice back to monocrystalline and cause the implanted dopant atoms to be substituted for silicon atoms in the silicon crystal structure. Without this heat cycle no transistor (or IC) action would occur.

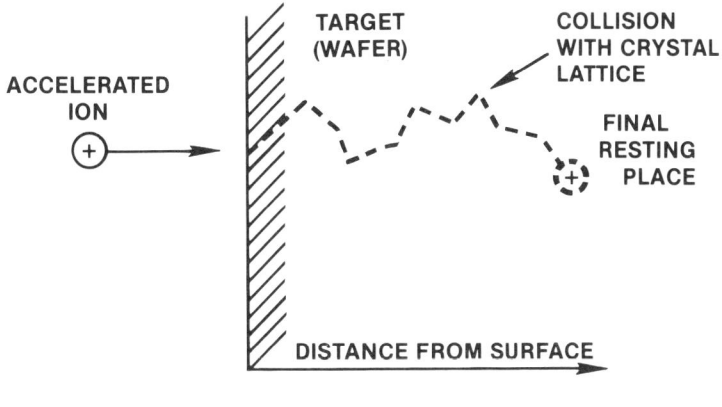

Figure 2-58. Path of an Implanted Ion

Evaluation of the implant process can be performed in a manner similar to that performed for the diffusion process. However, the test wafer used for sheet-resistance evaluation must be annealed before the sheet resistance can be measured.

Other evaluation alternatives are to use a special film on a glass disc and measure the optical properties of the film before and after implant, or use the therma-wave measurement system.

The therma-wave system measures the crystal damage of the implant and correlates the measurement to the implant dose and energy. This technique is the only measurement made immediately after the implant prior to the anneal. This provides immediate "feed-back" for process control.

INTEGRATED CIRCUIT ENGINEERING CORPORATION

The process engineer has more flexibility in masking ion implants than in diffusion. Diffusions are normally masked with oxides or oxide/nitride combinations. Implants can be masked by these, photoresists, and metal layers. Since implant is a low-temperature process, photoresist is a very common masking material. The engineer determines what thickness of resist is needed for a given implant species, dose, and energy. An example is shown in Figures 2-59 and 2-60.

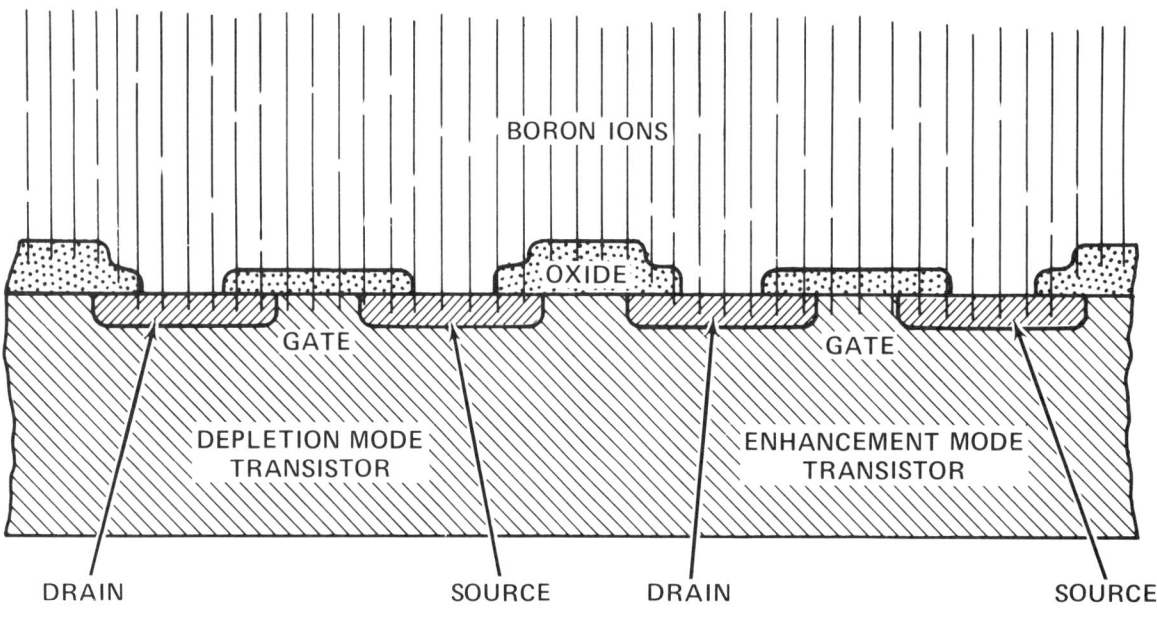

Figure 2-59. Step 1, The Adjustment of Thresholds

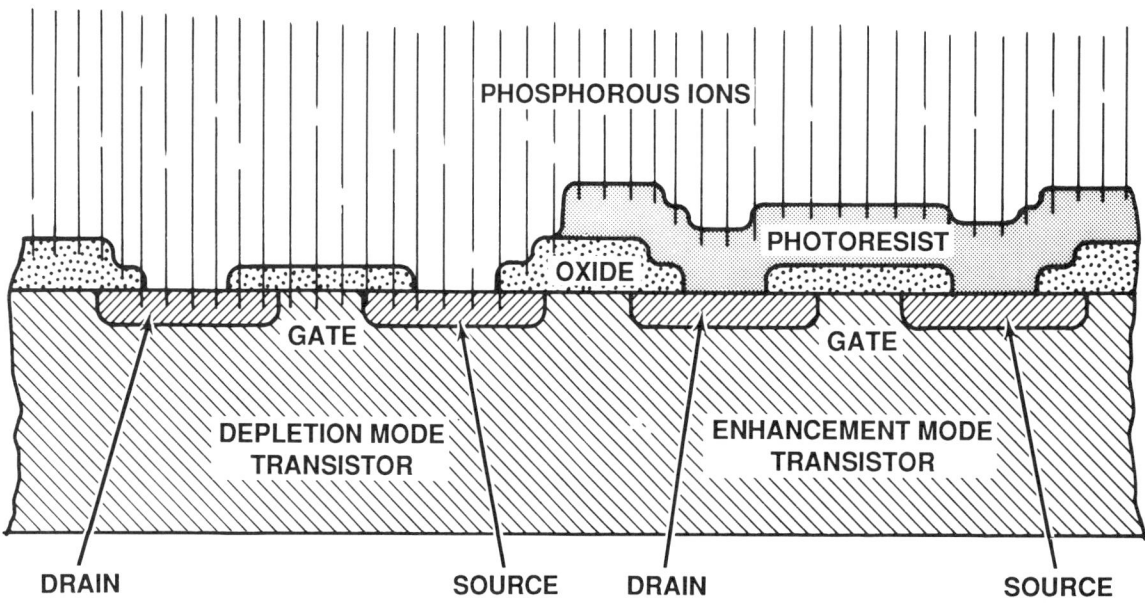

Figure 2-60. Step 2, Fabrication of the Depletion-Mode Transistors

Future uses of the implanter will expand it to other fab roles. In addition to the dopant role, the implanter can modify material characteristics, like gettering, etch rates, and grain boundaries to name a few.

F. EPITAXIAL DEPOSITION

The term epitaxial is derived from Greek, meaning to build upon. Epitaxial deposition, in general, is the deposition of a layer of single-crystal silicon on a single-crystal (monocrystalline)wafer. The deposited layer is a crystallographic extension of the substrate in terms of atomic order (i.e., it has the same crystal structure). Thus, the substrate could be considered the "seed" that is necessary to promote the single-crystal deposition.

Epitaxial deposition is a chemical vapor deposition (CVD) process. The subject of chemical vapor deposition was covered in the dielectric section (page 2-9). However, the original use of CVD started with the deposition of single-crystal silicon in the late 1950's and has played a major role in the industry since.

Epitaxy (or epi) has played a major role in the evolution of bipolar transistors and bipolar integrated circuits. In recent years both MOS discrete transistors and ICs have started to use epi as a key part of their structures.

Silicon CVD processes (epitaxial and polysilicon depositions) can be categorized by temperature range, pressure, and reactor wall temperature. The categories are summaries in Figure 2-61. In the case of polycrystalline silicon (poly, polysilicon) the crystallography of the substrate is not replicated during the deposition (e.g., poly on silicon dioxide). It is also possible that the deposition could yield an amorphorus film (e.g., silicon dioxide) that has no crystal structure. These types of films are shown in Figure 2-62.

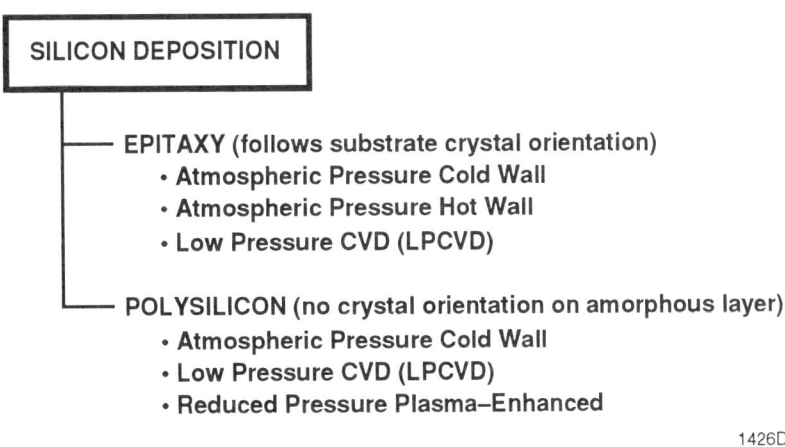

Figure 2-61. Silicon Chemical Vapor Deposition Processes

Basic Integrated Circuit Manufacturing

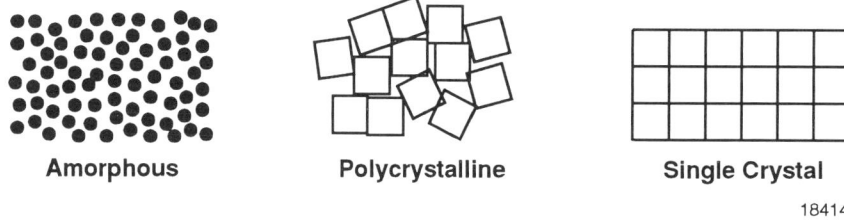

Figure 2-62. Types of Deposited Films

The primary purpose of the silicon epi process is to deposit high-quality, single-crystal films. During the deposition process dopant atoms are added to give the film the desired (N or P) electrical characteristics. The deposition time can be varied to control the thickness of the epi layer. The combination of the above provides the process with tremendous flexibility. Pictorial representations are shown in Figures 2-63 and 2-64.

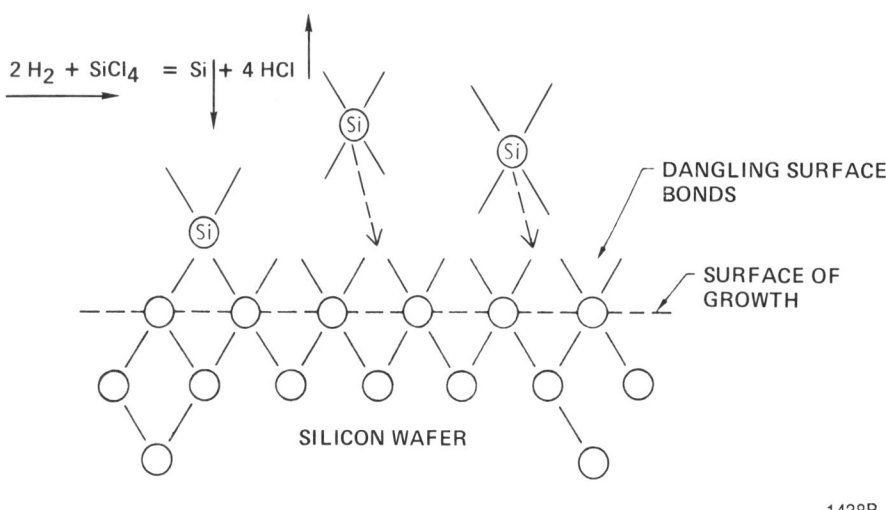

Figure 2-63. Pictorial Representation of Epitaxial Growth

Cleanliness of the epi process is critical to the quality of the film and the subsequent yields. This cleanliness must be considered in all aspects:

1. Wafer cleaning prior to the deposition process
2. Cleanliness (quality) of the chemicals used for the cleaning
3. Cleanliness of the gases going into the reaction chamber
4. Cleanliness of the inside of the reaction chamber
5. Cleanliness of the environment in the epi area
6. Cleanliness of the wafer handling.

Basic Integrated Circuit Manufacturing

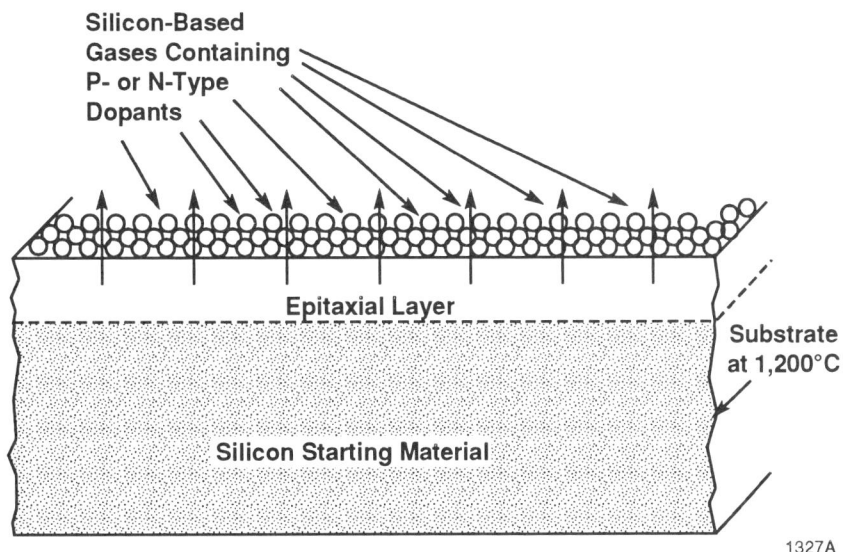

Figure 2-64. Epitaxial Growth

The evaluation of the epi layer measures the following.

1. Thickness
2. Resistivity
3. Surface quality
4. Crystallographic defects
5. Carrier lifetime
6. Dopant profile.

Thickness is measured with FTIR (Fourier Transform Infrared Spectroscopy) and resistivity is measured on a test wafer with the 4-pt. probe. Carrier lifetime and dopant profile are measured by C-V techniques. Surface quality and crystallographic defects are inspected by microscope.

The results of the epi process provide a wide range of epi thicknesses and resistivities on a variety of substrates. In addition, the deposited layer can be reasonably doped homogenously (evenly) as the deposition is done. For some devices this is an advantage over diffusion or implanting because it gives a graded doping.

The thickness can range from 0.6μm for high-performance devices to several hundred microns for discrete power products. The application of the product dictates the epi requirements.

G. POLYSILICON DEPOSITION

Polysilicon deposition is another application of CVD technology for thin films. Unlike epi, polysilicon is not required to follow the crystal orientation of the substrate (e.g., poly on silicon dioxide, which is amorphous).

Thin-film polysilicon has many important applications in the semiconductor industry. Polysilicon was the key ingredient leading to the "self aligned" MOS technology. Heavily-doped polysilicon has become the most widely used gate-electrode material for MOS products, both discretes and ICs. In addition, it serves as an interconnect, capacitor plate(s), doping source, and can be oxidized to form a stable layer of SiO_2. Polysilicon is utilized in these roles because of its compatibility with subsequent high-temperature processing, its excellent interface with SiO_2, its high reliability as a gate electrode material, and its ability to be deposited over steep topography with good conformal coverage. Lightly-doped polysilicon films are used as resistors in static memory products and to fill trenches in DRAMs.

Thin films of polysilicon are made up of small single-crystal grains of about 1000Å separated by grain boundaries. The film that will be a polysilicon layer can be either amorphorus or polycrystalline "as deposited." Subsequent heat cycles at elevated temperatures will cause an amorphous film to become polycrystalline. This "as-deposited" undoped film has an extremely high resistivity.

The resistivity of poly can be changed by doping the film with phorphous, boron, or arsenic. Because of the variation in grain size and grain boundary effects, heavily doping the film only reduces the sheet resistance to 10-30 $\Omega/\square$ (ohms/sq.). In between these extremes, ion implantation can control the resistivity for various resistor values or doping applications.

The deposition of polysilicon is generally done using LPCVD (Low-Pressure CVD) equipment operating in the 600-630°C range. This allows reasonable load size per run and acceptable productivity. Silane (SiH_4) is the silicon source material.

PECVD (Plasma-Enhanced CVD) reactors emerged in the late '80's and early '90's. PECVD systems can be operated at lower temperatures making them attractive for submicron technology. Figure 2-65 summarizes LPCVD deposition. Figure 2-66 illustrates the concept of the system.

- **LPCVD MID-TEMPERATURE (600 - 700°C)**
  — Vertical-Stacked Wafers
  — Resistance Heating
  — Pressure (100 - 500 Millitorr)
  — In-Situ Doping
- **PECVD LOW-TEMPERATURE (LESS THAN 300°C)**
  — Horizontal Wafer Position
  — Vertical-Stacked Wafers
- **ADVANTAGES OF LOW PRESSURE PROCESSING**
  — Uniformity
  — Lower Temperature
  — Less Wafer Warpage
  — Lower Cost
- **DISADVANTAGES**
  — Pyrophoric Gas
  — Toxic Doping Gases
  — Exhaust Scrubbing

Figure 2-65. Low-Pressure Polysilicon Deposition

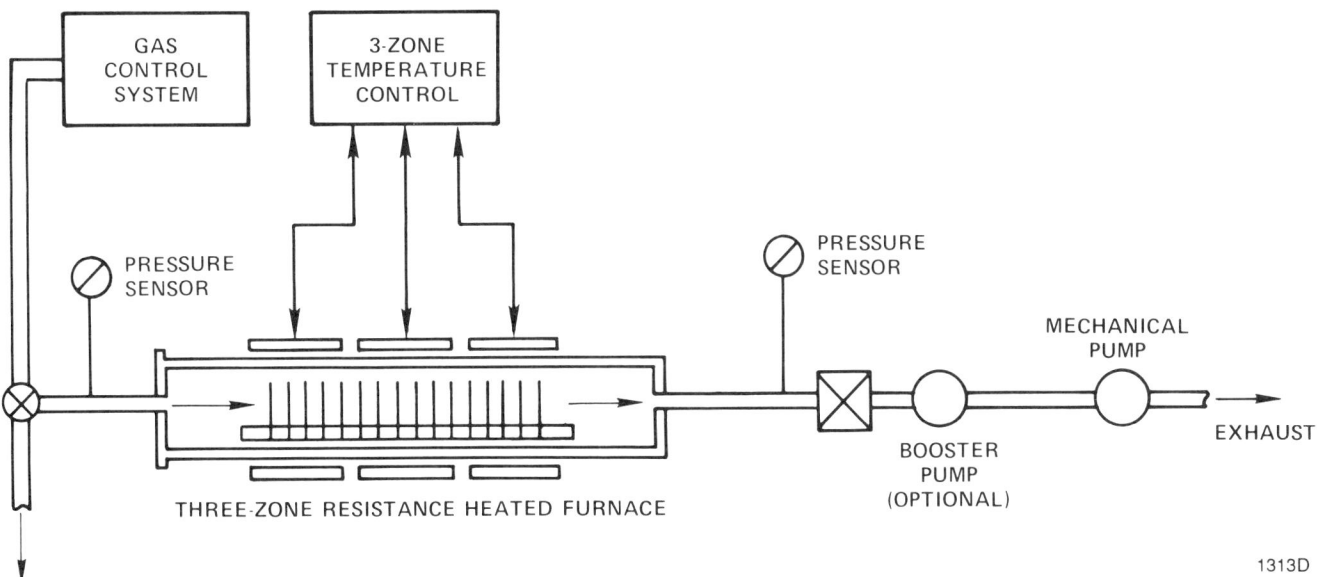

Figure 2-66. Block Diagram of a Low-Pressure Chemical Vapor Deposition System

Earlier, it was indicated the polysilicon could be doped to control the resistivity. The doping process for polysilicon can be a separate process after the deposition or it can be incorporated into the deposition. There are advantages and limitations to each process technique.

H. METAL DEPOSITION

Metal (usually aluminum or aluminum with a small amount of other materials) is used to connect the individual components (diodes, transistors, resistors, and capacitors) in an integrated circuit. A layer of metal is deposited on the entire top surface of a wafer. Through photolithography and etch, selected portions of the metal are removed. The remaining aluminum serves as the conductors (wires) between the various components of each IC.

Noyce's revolutionary concept of wiring all the transistors at the top surface of the silicon rather than having individual discrete dice on a substrate altered the direction of the semiconductor industry. The prior art of the Kilby patent integrated components on a common substrate but still used loops of very fine wire to interconnect the components. Noyce's new method of wiring circuits all within the die area of the circuit opened up endless new possibilities. This section will cover the metal technology used for interconnects.

Basic Integrated Circuit Manufacturing

1. Process Sequence

The metallization process sequence is shown in Figure 2-67. This figure illustrates that an earlier photo-patterning process has removed the silicon dioxide from the areas where ohmic contact will be required. A layer of metal is deposited over the surface and the photo-patterning process is repeated to delineate the metal interconnect. This simplification of the metallization process shows the genius of Noyce. The process concept is the same for interconnecting two transistors or seventy million transistors. Further, this fundamental concept can be extended to more than one level of interconnect as will be shown later.

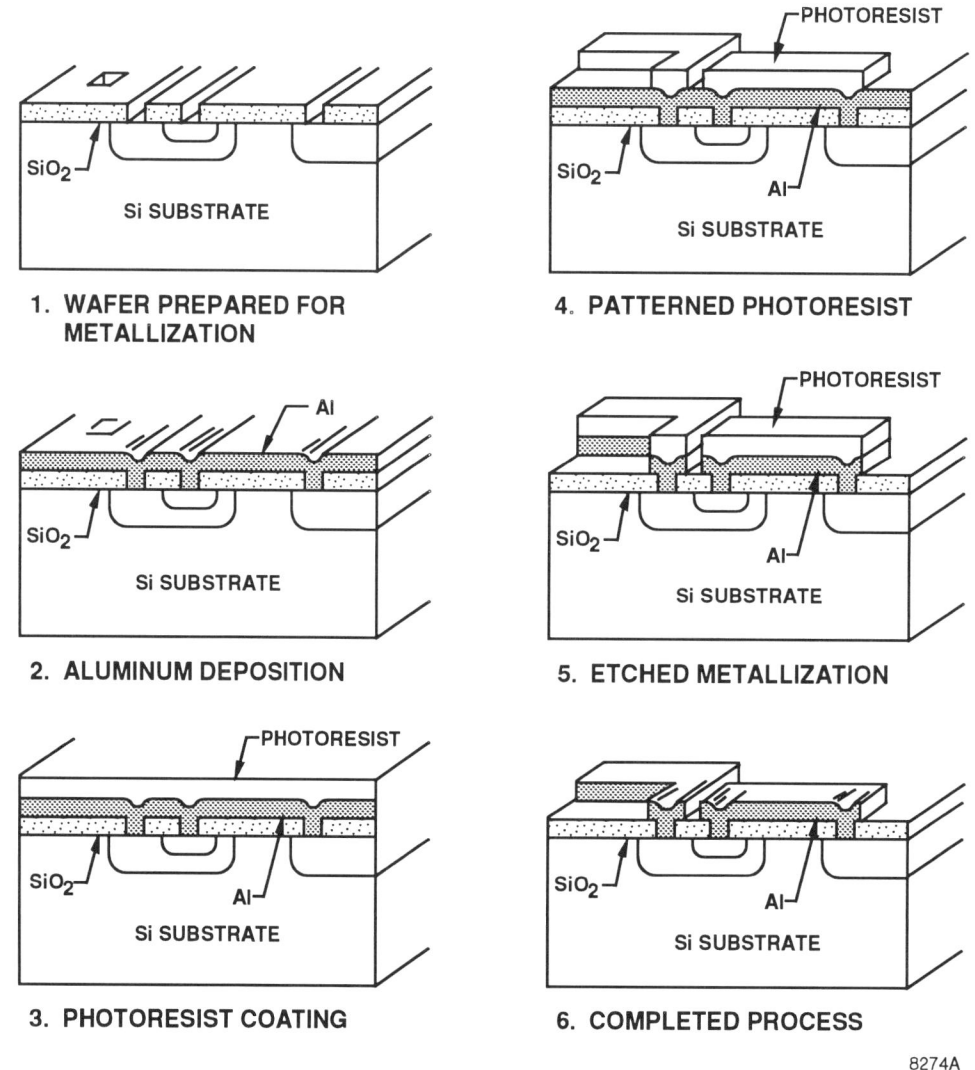

Figure 2-67. Metallization Process Sequence

Basic Integrated Circuit Manufacturing

The metal deposition process flow is shown in Figure 2-68. The wafer(s) receive a wet chemical clean just prior to loading into the deposition system. The deposition system is equipped with a vacuum system for controlling the internal partial pressure and system electronics to control the deposition process. The resulting film is measured for film thickness, uniformity, and step coverage. Figure 2-69 diagrams a metallization system.

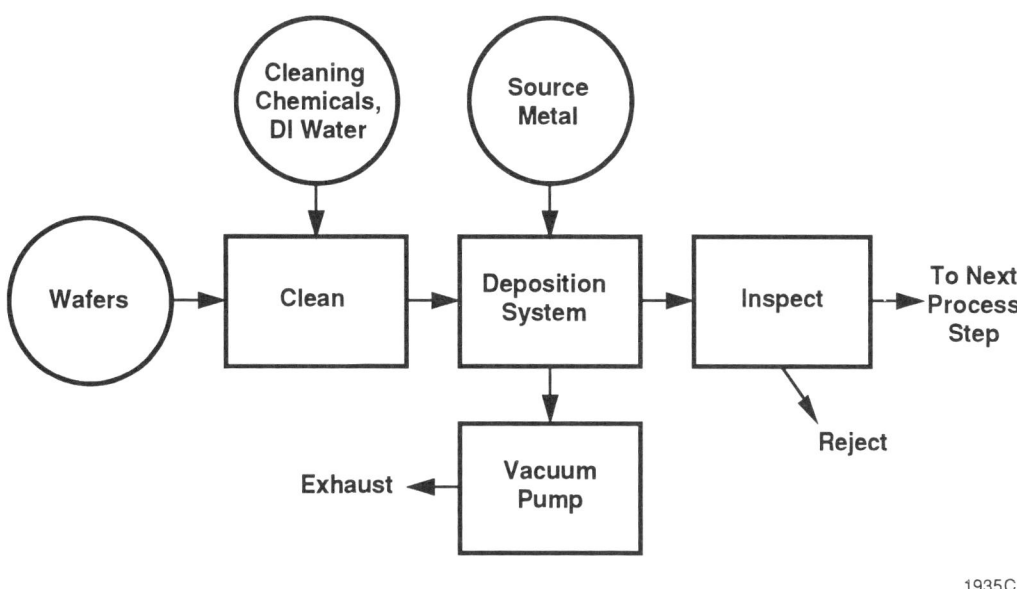

Figure 2-68. Metal Deposition Process Flow Chart

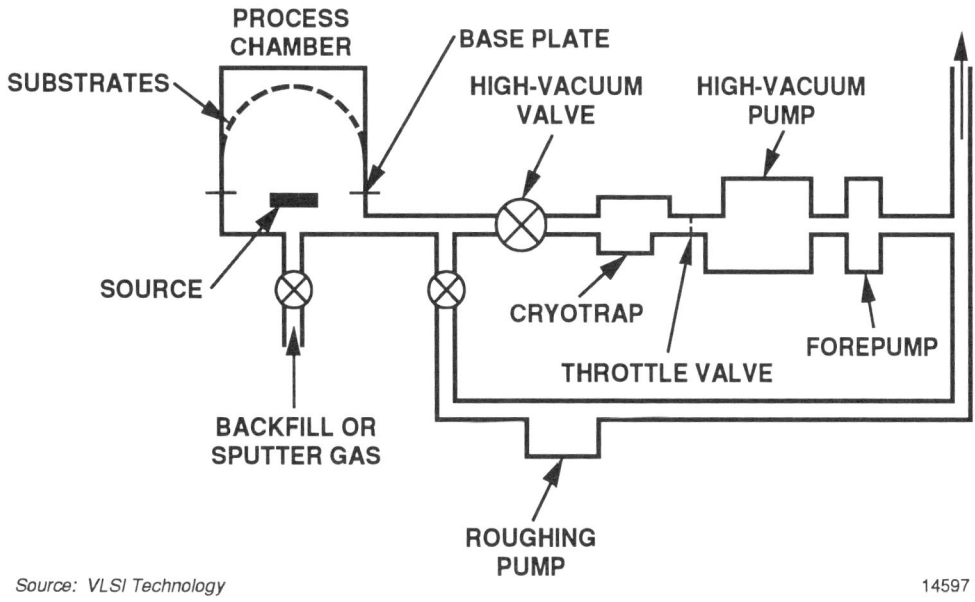

Figure 2-69. Metallization System

Basic Integrated Circuit Manufacturing

- Low electrical resistivity
- Ohmic and low contact resistance
- Stable contact formation to silicon or other metal(s)
- High temperature stability
- Excellent adhesion and low stress
- Good electromigration resistance
- Good corrosion resistance
- Controlled oxidation properties and stability in an oxidizing environment
- Ease of formation
- Ease of fine line pattern transfer
- Smooth surface features

Figure 2-70. Conductor Material Properties Requirements for VLSI

2. Materials

The basic materials requirements to interconnect the components within an IC are listed in Figure 2-70. Unfortunately, no individual metal can meet all of the requirements. Aluminum is the most often used metal but its use is being challenged by newer circuit requirements.

When pure aluminum is used to interconnect shallow p-n junction devices, a reliability problem can occur. The silicon atoms in the substrate diffuse into the aluminum metallization. Over extended time and temperature, sufficient silicon is depleted to cause a metal short through the p-n junction. This concept is illustrated in Figure 2-71. The problem can be minimized by adding a small percentage of silicon into the aluminum.

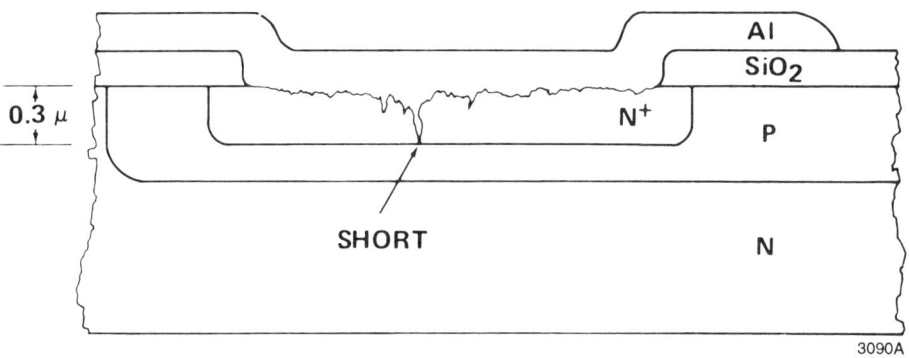

Figure 2-71. Aluminum/Silicon Dissolution

3. Methods

a. E-beam, Filament Evaporation

The methods for depositing metal films are tabulated in Figure 2-72. The semiconductor industry has evolved through each of the methods except Ion Beam deposition, which has not been implemented to any great extent yet. Sputtering is the most commonly used method while LPCVD/PECVD is being used for selected films of tungsten. Diagrams for filament evaporation and electron-beam (e-beam) are shown in Figure 2-73 and Figure 2-74, respectively.

Figure 2-72. Methods of Metal Deposition

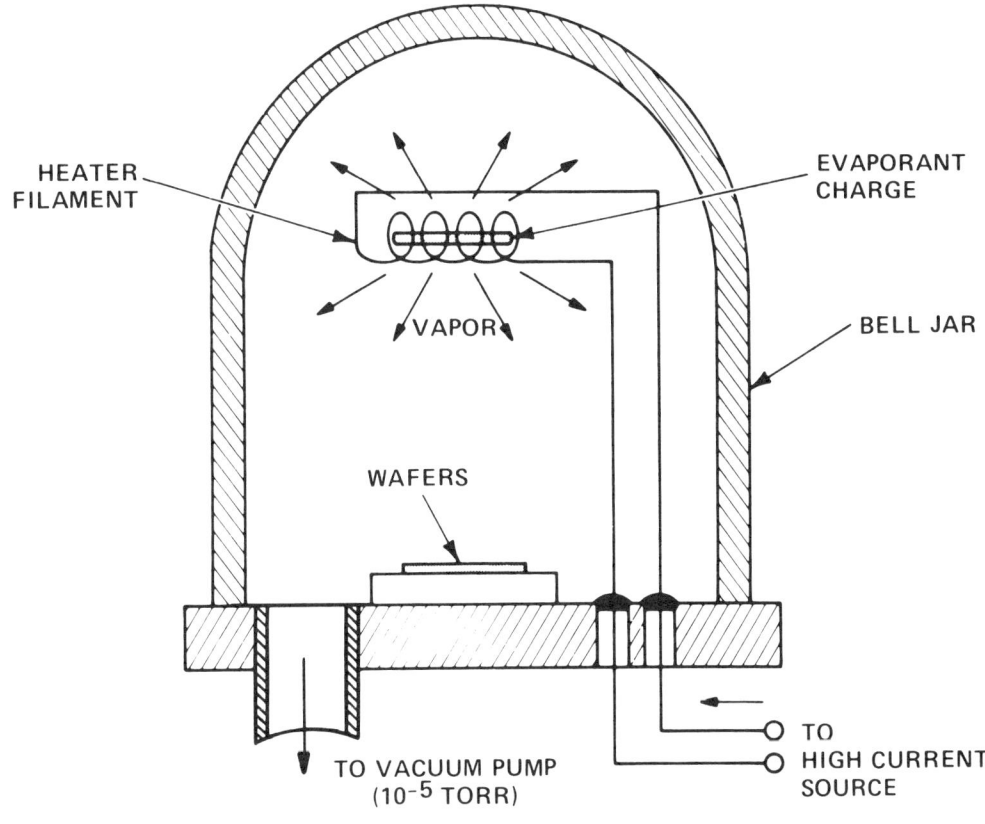

Figure 2-73. Vacuum Evaporation System

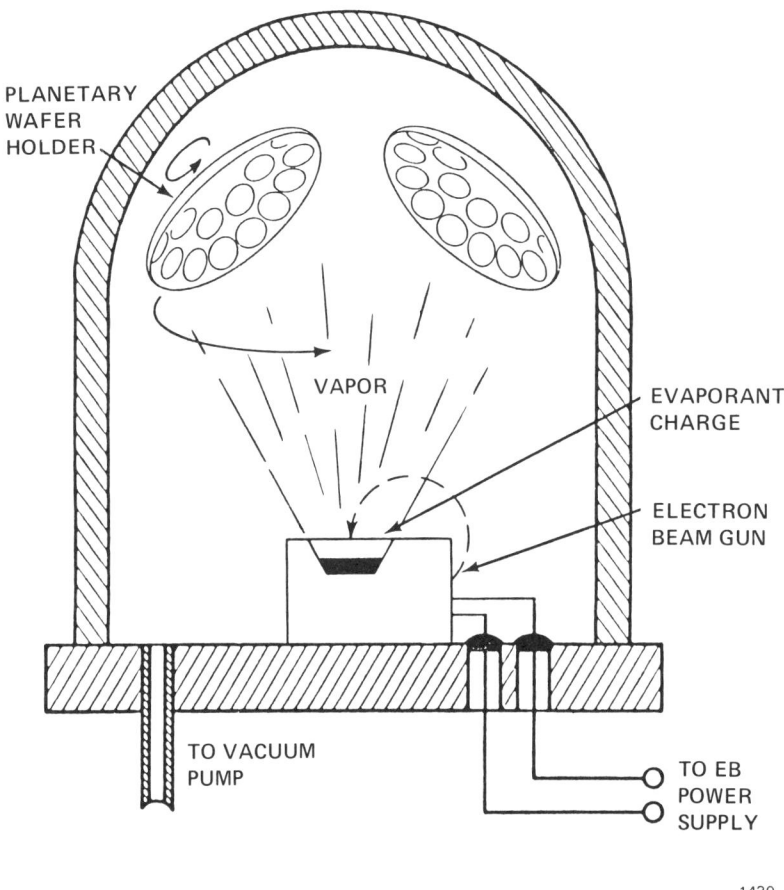

Figure 2-74. Electron Beam Evaporation

The electron-beam system requires a separate e-gun for each material to co-deposit the film. This increases the cost of the deposition system.

b. Sputtering

Low-pressure sputtering evolved as an improved alternative to the e-beam deposition system. A sputtering system is diagrammed in Figure 2-75. The main advantage of sputtering is the physical nature of the process. Argon is the ionized atom most commonly used to bombard the target material. This physical bombardment causes the target material to be deposited from many different angles, which provides a more uniform deposition, regardless of target composition. The resulting step coverage is improved as shown in Figure 2-76.

Many refinements have been made on the hardware of sputtering deposition systems. These improvements have made it the method of choice for the majority of wafer fabs. Various ways of configuring the sputtering system were noted in Figure 2-72. One of the more popular concepts is shown in Figure 2-77.

Basic Integrated Circuit Manufacturing

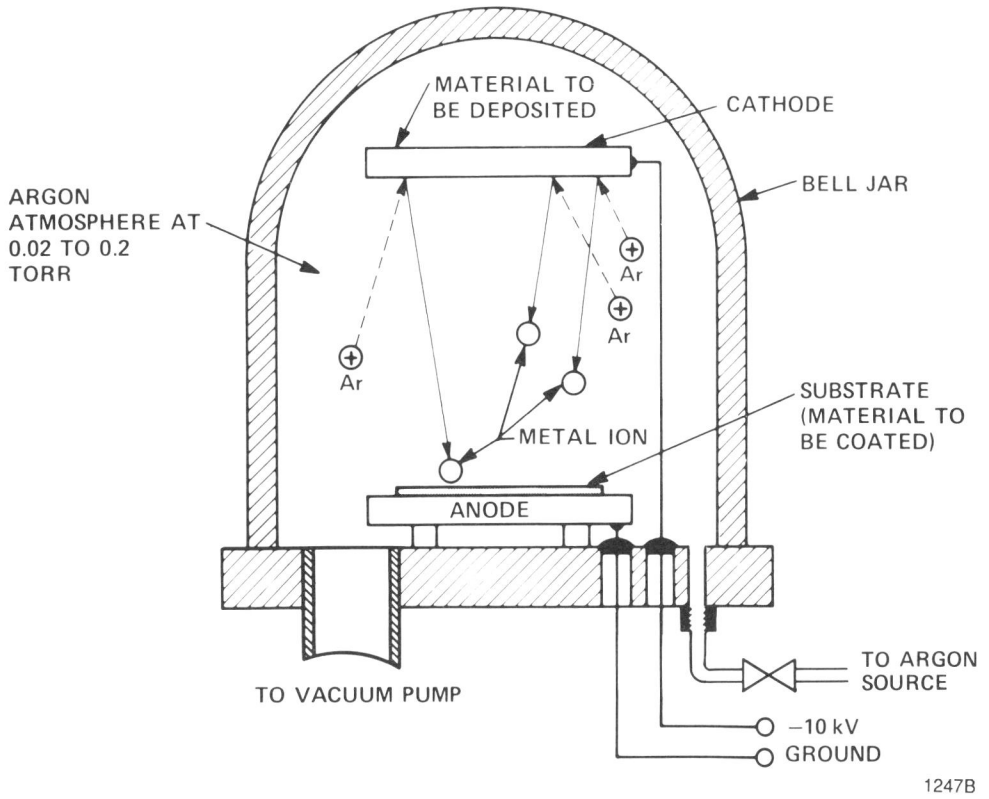

Figure 2-75. Low-Pressure Sputtering

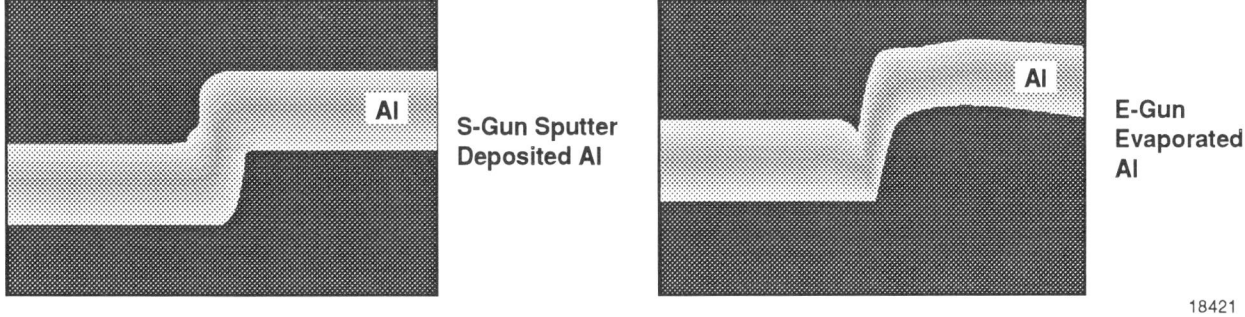

Figure 2-76. Aluminum Step Coverage Comparison

Basic Integrated Circuit Manufacturing

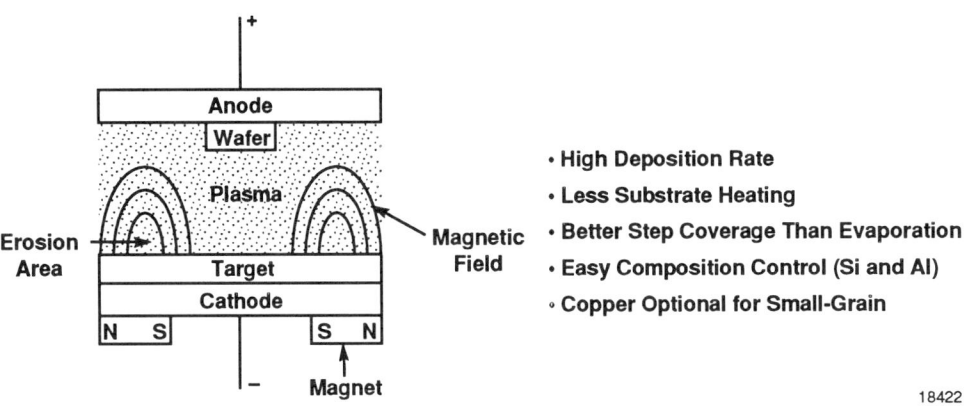

Figure 2-77. Planar DC Magnetron Sputtering

4. Electromigration

Another reliability problem associated with the interconnect metallization is electromigration (similar to blowing a fuse, see glossary for definition). The details are beyond the scope of this book but the general concept is illustrated in Figure 2-78. From the drawing it can be observed the problem is sensitive to the thickness of the film. Thickness uniformity over topological changes is known as step coverage. Poor step coverage contributes to the problem of electromigration. To further reduce electromigration, a small percentage of copper is added to the silicon-aluminum metal. Semiconductor manufacturers evaluate their products on an "on-going" basis for this characteristic.

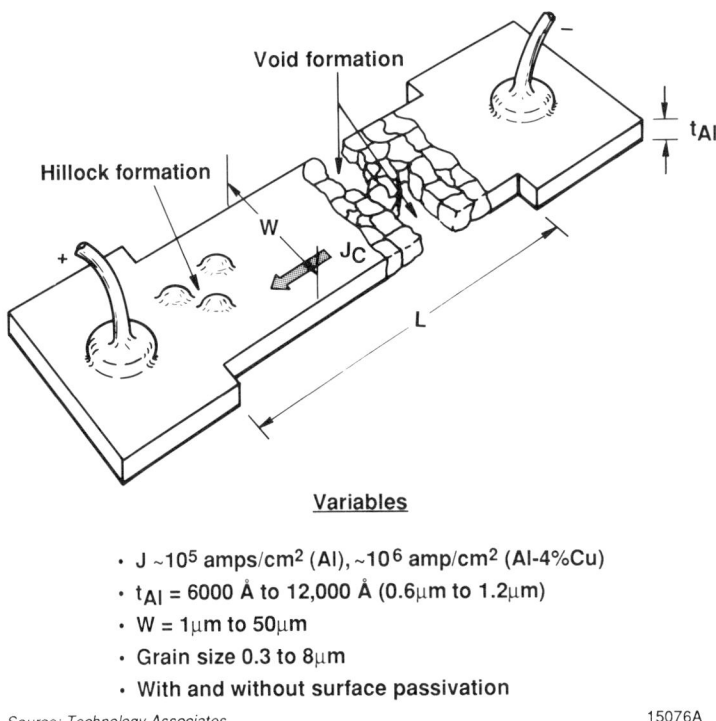

Variables

- $J \sim 10^5$ amps/cm^2 (Al), $\sim 10^6$ amp/cm^2 (Al-4%Cu)
- t_{Al} = 6000 Å to 12,000 Å (0.6μm to 1.2μm)
- W = 1μm to 50μm
- Grain size 0.3 to 8μm
- With and without surface passivation

Source: Technology Associates

Figure 2-78. Electromigration Description

5. Alloy (Sinter)

The successful patterning of the interconnect metal does not guarantee that adequate ohmic contact has been made to the silicon. An "alloy" or "sintering" process step is usually used to insure a low-resistance contact between the interconnect metal and the silicon. The alloy step is done in a furnace at a temperature in the range of 400-475°C. The alloy temperature, time and ambient will vary from process to process, but the limits are determined by the metal composition and junction depth considerations in conjunction with the metallurgical phase-diagram. Sometimes this alloy step is referred to as an anneal step.

6. Multilayer Interconnect

As the integration density of ICs has increased, interconnecting the various components has an impact on the die size (interconnect takes up space). To realize the maximum number of interconnects per unit area, multiple layers of interconnects are used. This is illustrated in Figure 2-79.

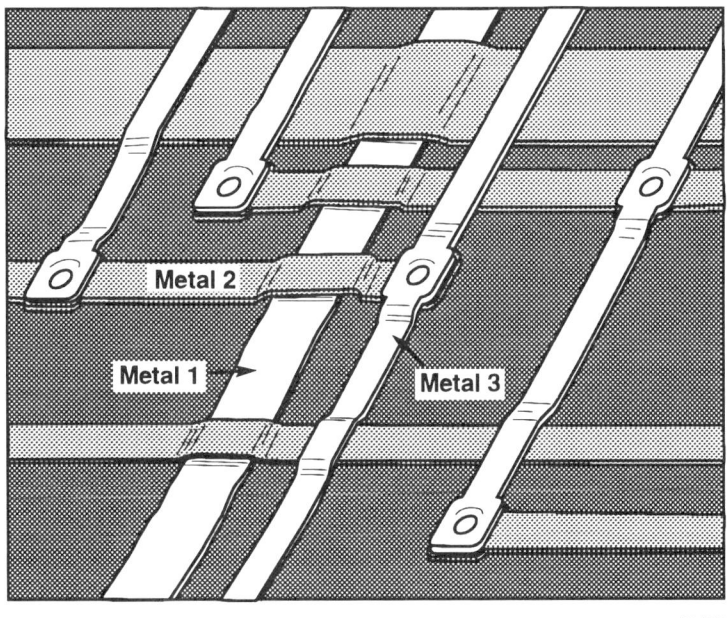

Figure 2-79. Three-Layer Metallization on IC Chip

Multilayer interconnect has placed new demands on the metallization capabilities of existing processes. Research has produced several alternative materials, processes, and equipment.

a. Barrier Metals

The use of a barrier metal at the silicon-to-metal and metal-to-metal interfaces has reduced the junction shorting/leakage problem and improved ohmic contact resistance consistency. Barrier metals are also used on top of the metal layers to assist the various considerations in patterning and etch. Further, the barrier improves the interface between layer 1 and layer 2, layer 2 and layer 3, etc. A barrier metal is illustrated in Figure 2-80.

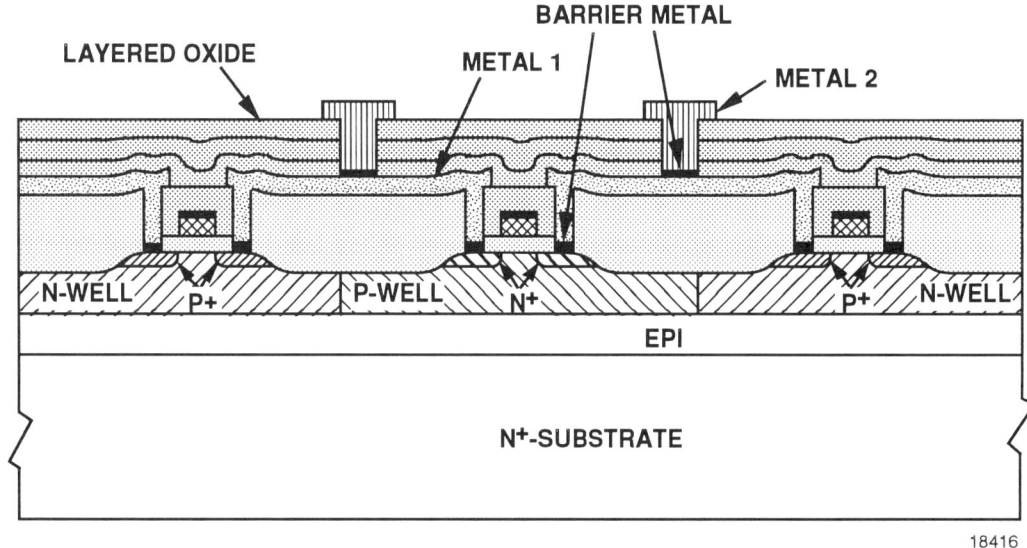

Figure 2-80. Double-Layer Metal With Barriers

b. Vias

In conjunction with barrier metal technology, the capability to fill an ohmic contact area with another metal was developed. Frequently this fill metal is tungsten. Tungsten can be deposited to conform to the topological changes and it is easy to dry etch. This is illustrated in Figures 2-81 and 2-82.

c. Dielectric Isolation

Another part of the multilayer metal technology is the dielectric material used to separate the layers of metallizations. Various types of CVD and spin-on dielectrics are used. To smooth out the topography of the surface of these dielectrics, various etch backs and mechanical polishing are used.

Generally, after the first layer of metal has been completed, the "alloy" or "sintering" process step is done and the wafers' test patterns are tested.

Basic Integrated Circuit Manufacturing

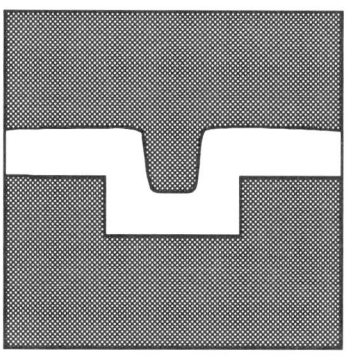

Tungsten First-Level-Metal
- Electromigration and hillock elimination.
- Greater than 95% conformality.
- Low contact resistance to N+ silicon.
- High temperature processing.
- Planarizes contacts.

Source: Genus

Figure 2-81. CVD Tungsten Thin Film

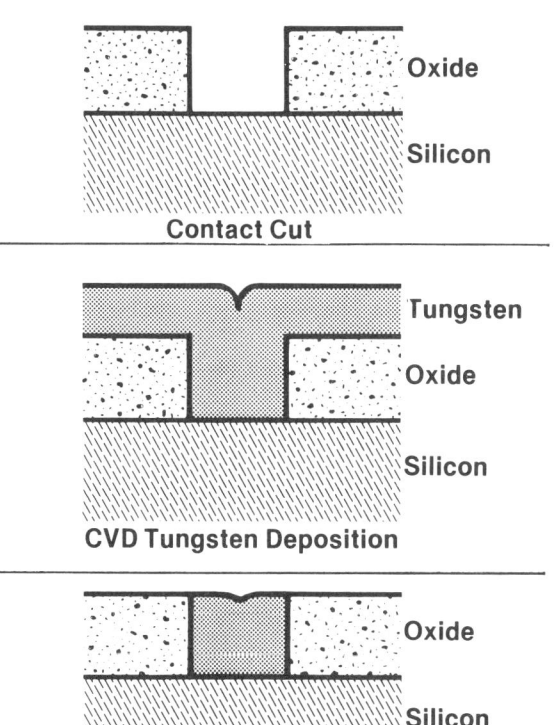

TUNGSTEN PLUG TO SILICON

- Low contact resistance to N$^+$ silicon.

- Planarizes contacts.

- Slope-etch not required.

- Eliminates aluminum Physical Vapor Deposition step coverage problems into vias.

- Easily etchable using $CF_4 + O_2$, SF_6

Source: Genus

Figure 2-82. For Via Fill to Silicon

Basic Integrated Circuit Manufacturing

I. PROCESSES

1. CMOS

Figures 2-83, 2-84, and 2-85 provide a graphical representation of the IC manufacturing cycle. Figures 2-86, 2-87, and 2-88 provide graphical cross-sections of the CMOS structure as the circuit travels through the manufacturing cycle. Each group of process steps are narrated relative to Figures 2-83, 2-84, and 2-85.

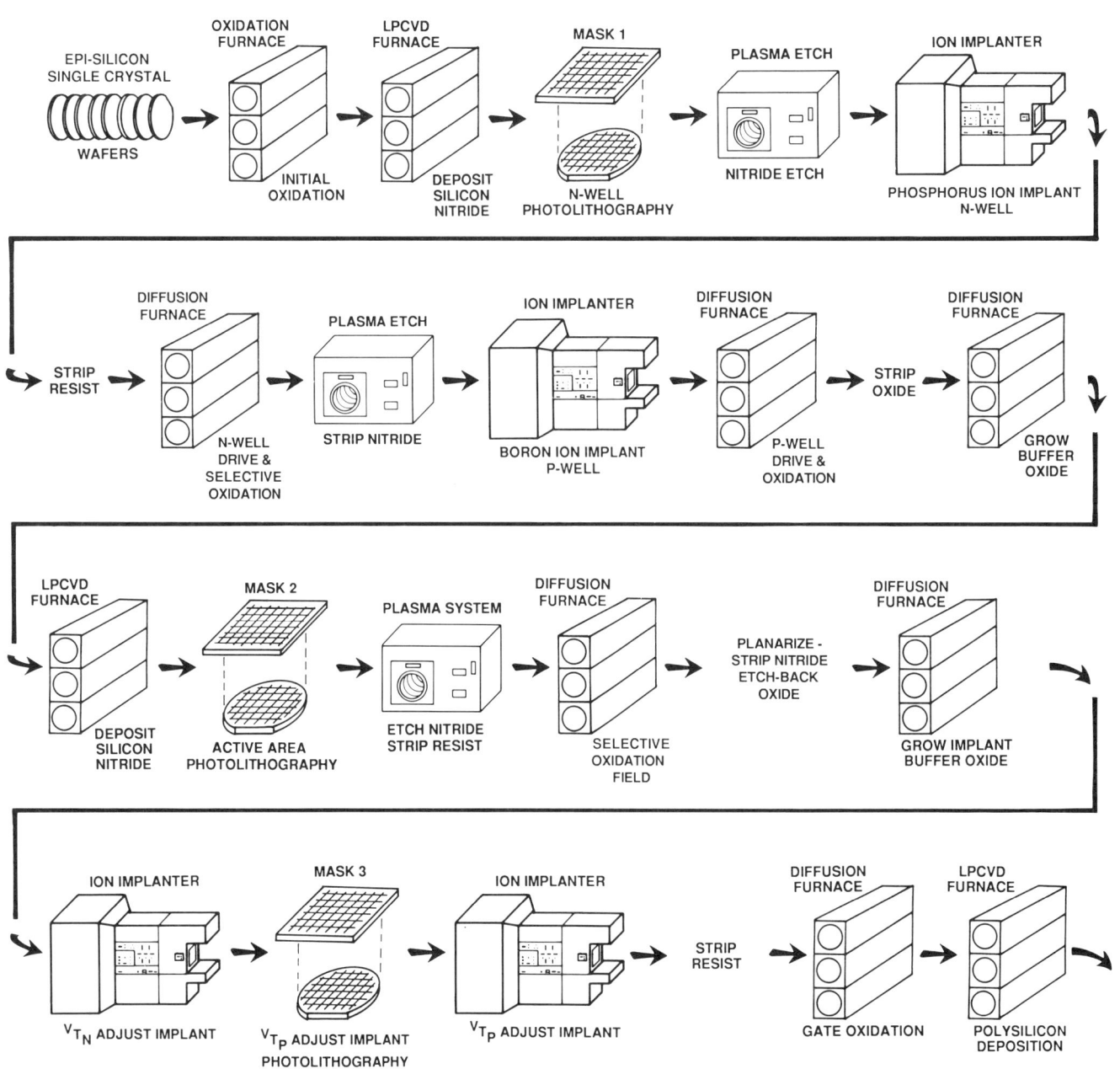

Figure 2-83. Twin-Well Silicon-Gate CMOS Manufacturing Sequence (1 of 3)

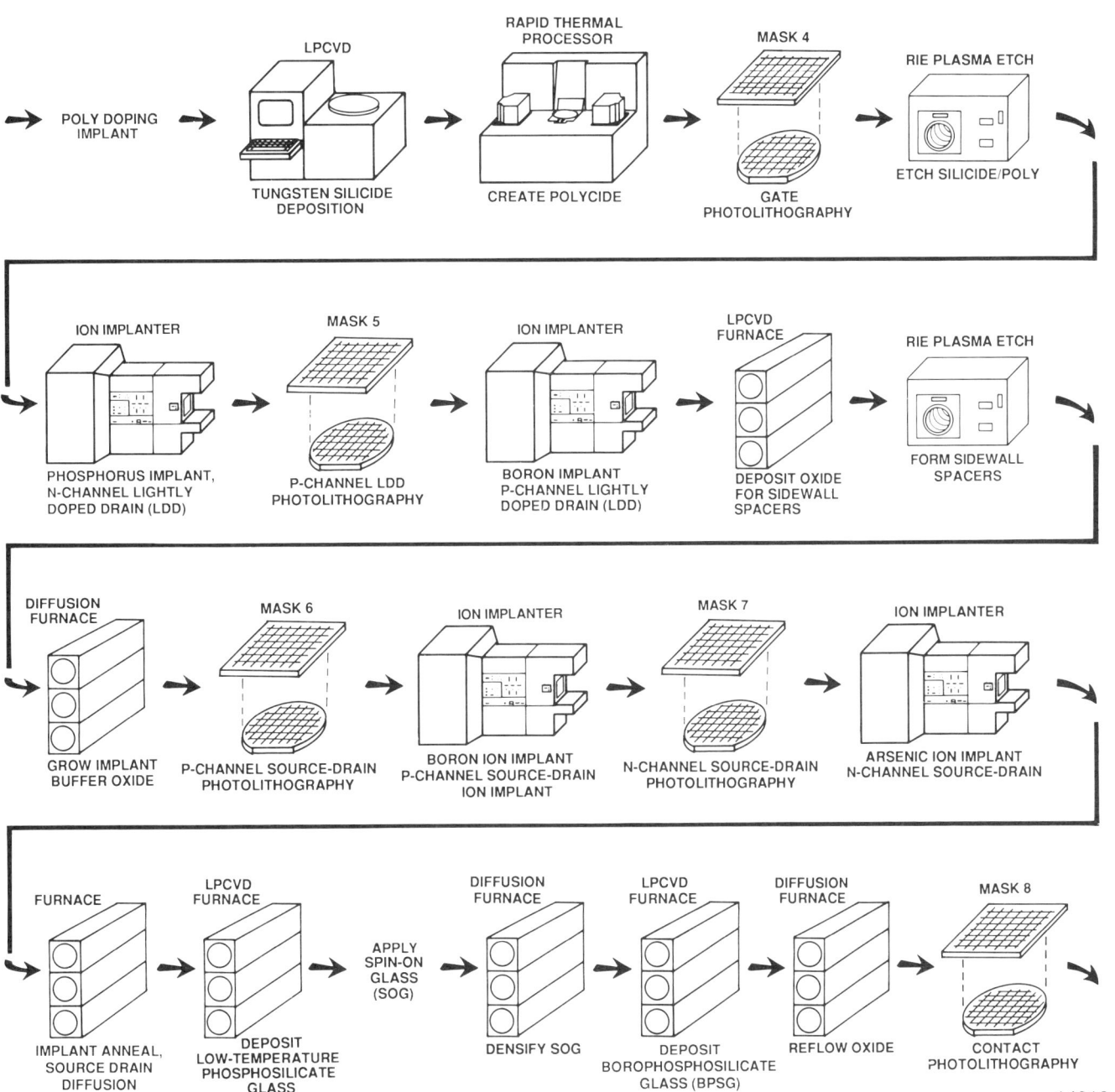

Figure 2-84. Twin-Well Silicon-Gate CMOS Manufacturing Sequence (2 of 3)

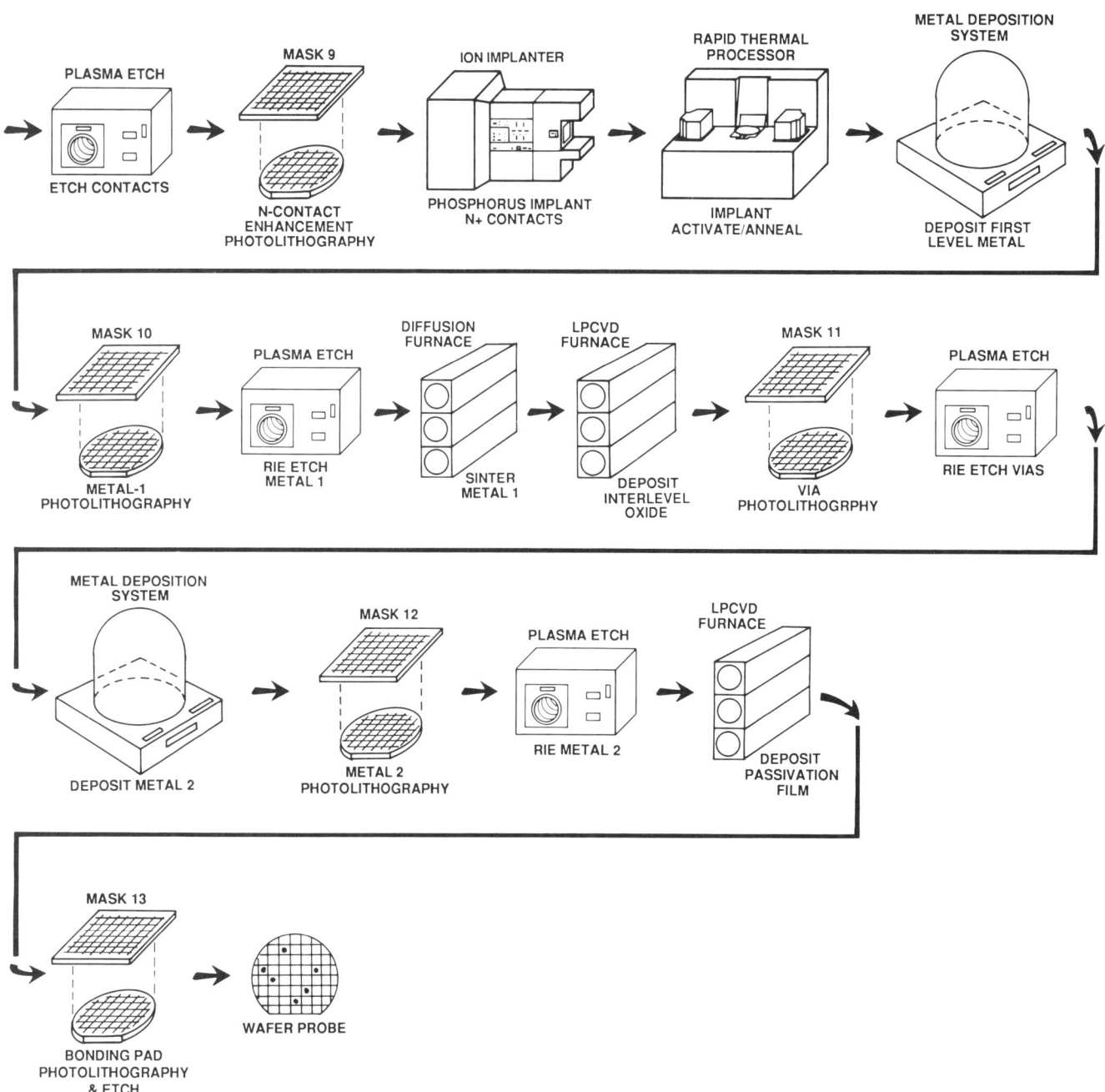

Figure 2-85. Twin-Well Silicon-Gate CMOS Manufacturing Sequence (3 of 3)

Basic Integrated Circuit Manufacturing

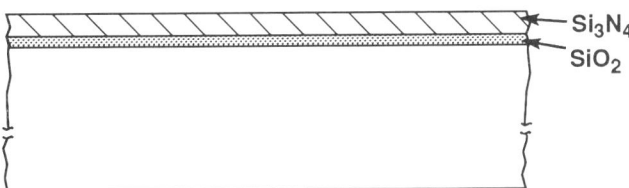

1. Starting material with SiO$_2$ and Si$_3$N$_4$.

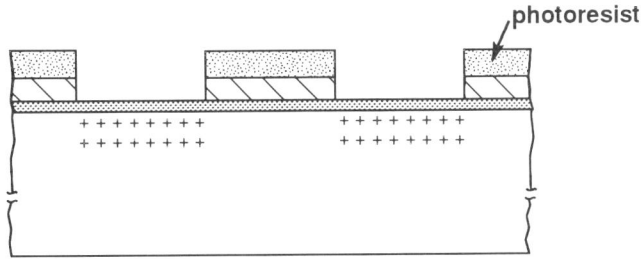

2. N-well lithography, etch, phosphorus implant, resist strip, N-well drive and oxidation, strip nitride

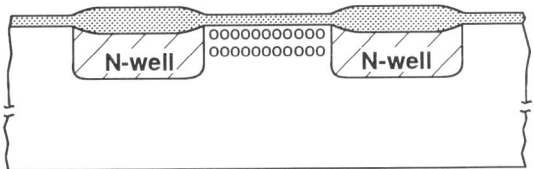

3. Boron P-well implant, P-well drive/oxidation, strip all oxide, grow thin oxide and deposit nitride

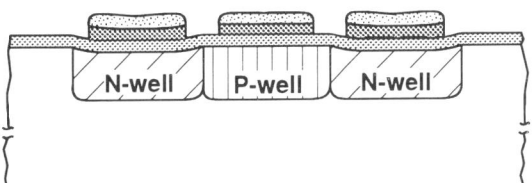

4. Active area lithography and etch nitride, strip photoresist

17950

Figure 2-86. Twin-Well Silicon-Gate CMOS Process Flow (1 of 3)

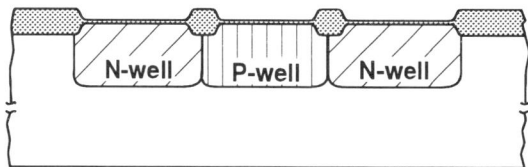

5. Grow field oxide, strip nitride, oxide etch back, grow thin oxide

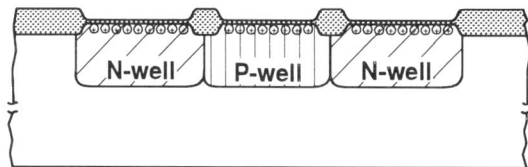

6. V_{TN} lithography, V_{TN} implant, strip resist, V_{TP} lithography, V_{TP} implant strip resist, RTA implants, strip thin oxide

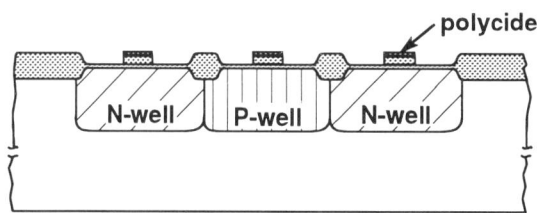

7. Grow gate oxide, deposit polysilicon, implant polysilicon with arsenic, deposit tungsten silicide, anneal, gate lithography, etch polycide, strip resist

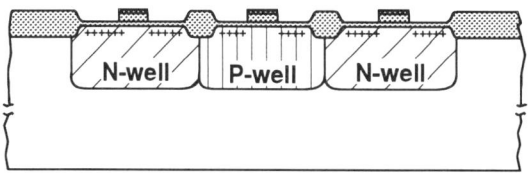

8. LDD N-channel lithography, LDD N-implant, strip resist, LDD P-channel lithography, LDD P-implant, RTA implants

Figure 2-87. Twin-Well Silicon-Gate CMOS Process Flow (2 of 3)

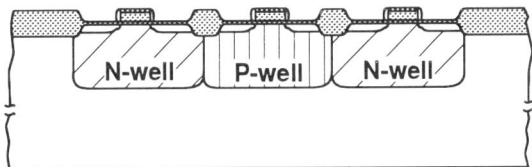

9. Deposit spacer oxide, RIE etch back spacer oxide,
N+ S/D lithography, N+ S/D implant, strip resist,
P+ lithography, P+ S/D implant, strip resist, anneal implants

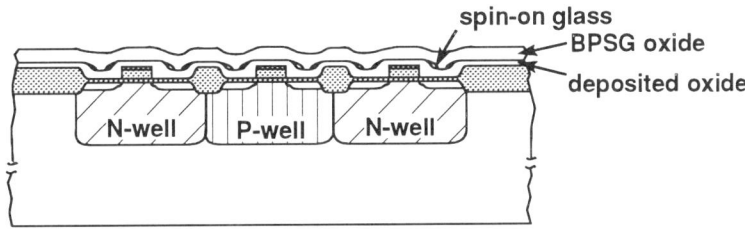

10. Deposit oxide, apply spin-on glass, deposit
BPSG oxide, reflow oxides, etch back reflow oxide

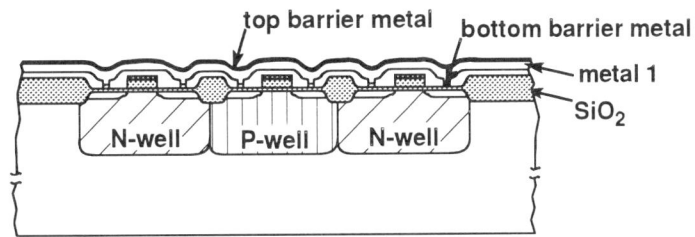

11. Contact lithography, bottom barrier metal deposition,
aluminum deposition, top barrier metal deposition

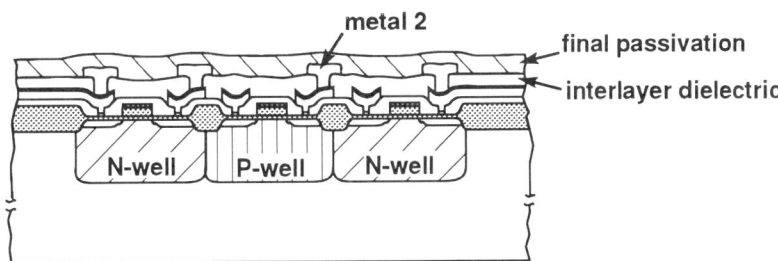

12. Metal 1 lithography, deposit interlayer dielectric,
via lithography, deposit metal 2, metal 2 lithography,
anneal, deposit final passivation

17952

Figure 2-88. Twin-Well Silicon-Gate CMOS Process Flow (3 of 3)

2. NMOS

A simple NMOS IC process is shown in Figure 2-89.

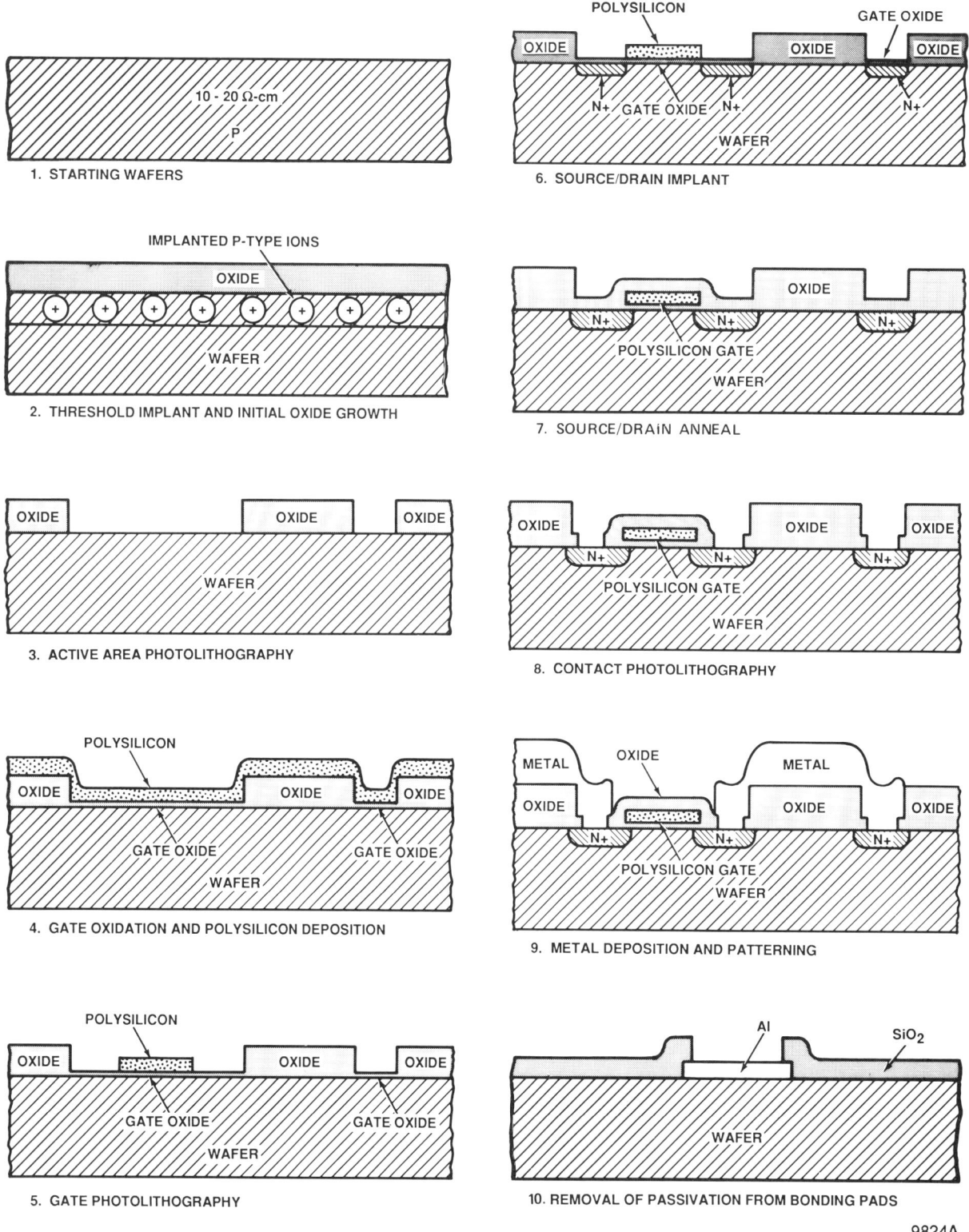

Figure 2-89. NMOS Process

Basic Integrated Circuit Manufacturing

3. Bipolar

A simple bipolar process is shown in Figure 2-90.

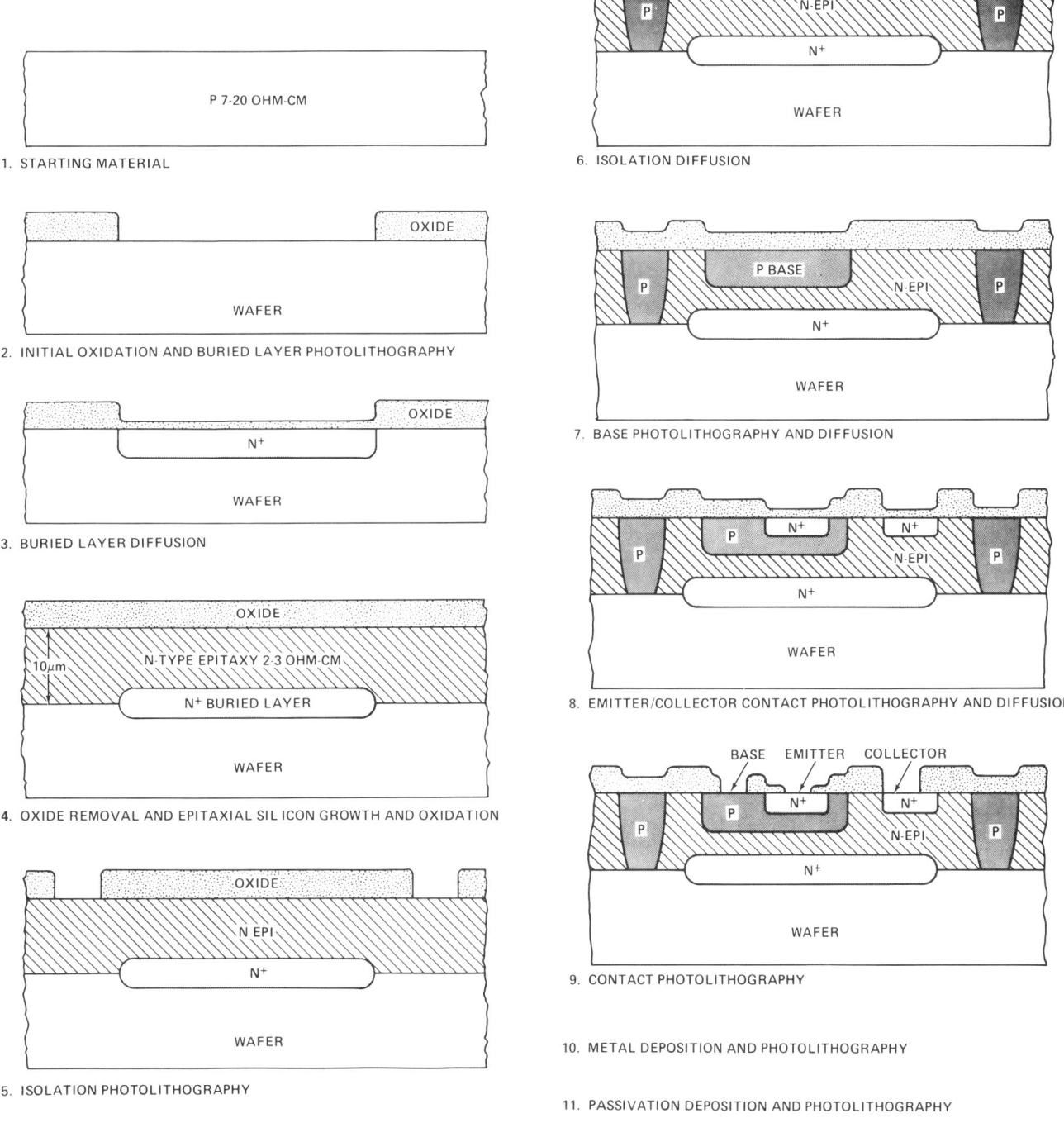

Figure 2-90. Bipolar Process Flow

4. Bipolar (ECL)

An advanced Bipolar ECL IC process is shown in Figure 2-91.

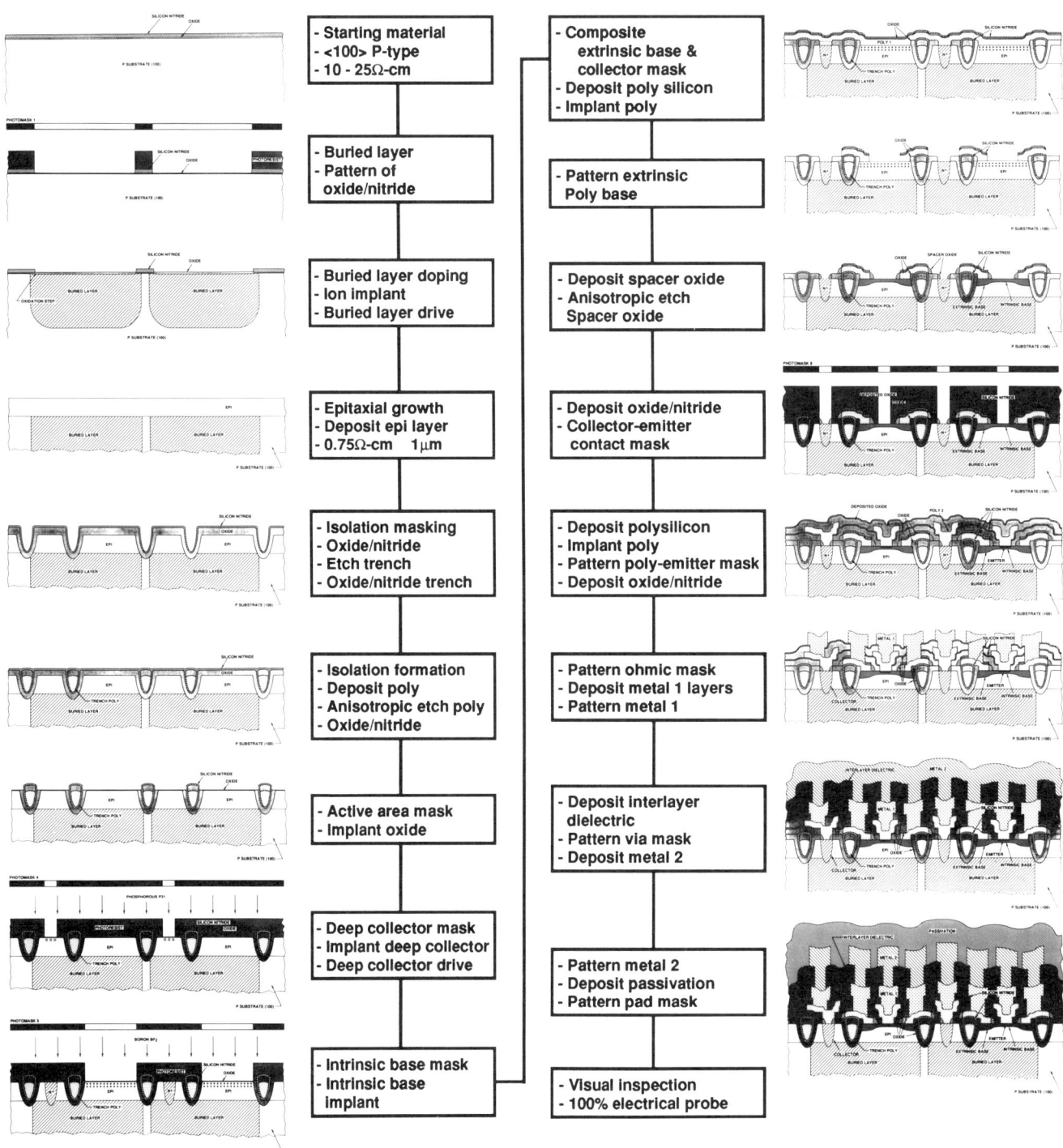

Figure 2-91. Advanced Bipolar Digital Process Summary

J. ASSEMBLY

After all the wafer fabrication steps have been completed, the good dice within the wafer are ready to be separated and assembled into final product form. The assembly or packaging process will place the electrically good devices (chips, die, and dice are other terms used to identify the individual circuits) in a package, interconnect the device to the package's leads and provide some form of final sealing.

Since the early days of the semiconductor industry, the packaging process has often been considered to be a lower level of technology than the wafer fab process. As to why this label emerged is purely speculation, but because of the high-labor content per package, the volume assembly of semiconductors was relocated to the Asia-Pacific region. This region continues to dominate the volume assembly.

As the semiconductor industry moved from the SSI (small-scale integration) era through MSI and LSI (medium and large) into the VLSI (very large-scale integration) levels of integration, packaging technology has changed dramatically. The changes have included automation, advances in materials, and new package designs.

The IC assembly sequence is outlined in Figure 2-92. The wafer is sawn into individual dice (separated). Each good die is attached to a package substrate (as shown) or leadframe. The bonding pads around the perimeter of the die are connected (usually wire bonded) to the internal terminations of the external leads. Finally, the package is completed by sealing the pieces of the housing together or by encapsulation with molding compound. This sequence will convert a tested wafer into a packaged product. The basic concepts of packaging are illustrated in Figure 2-93. Packaging requirements can be translated into package functions and constraints, which are shown in Figure 2-94. These functions and constraints along with the package design factors dictate the actual package design. The design factors consider the chip performance requirements, the density of chip, materials issues, assembly methods, and cost. These factors are expanded in Figure 2-95.

The cost factor has played a major influence on packaging. As previously stated, labor cost was responsible for assembly moving to the Asia-Pacific region. Material costs for plastic packages are less expensive than for ceramic or metal packages. Plastic package assembly methods have been automated to a higher level than either ceramic or metal packages. All of these cost factors favor the plastic package and have elevated it to be the highest volume type of IC package produced.

The major segments in the assembly process are outlined in Figure 2-96. These generic outlines can be further divided into flows for plastic packages and ceramic packages. These flows will be illustrated after reviewing the die separation, die bonding, and wire bonding processes.

Basic Integrated Circuit Manufacturing

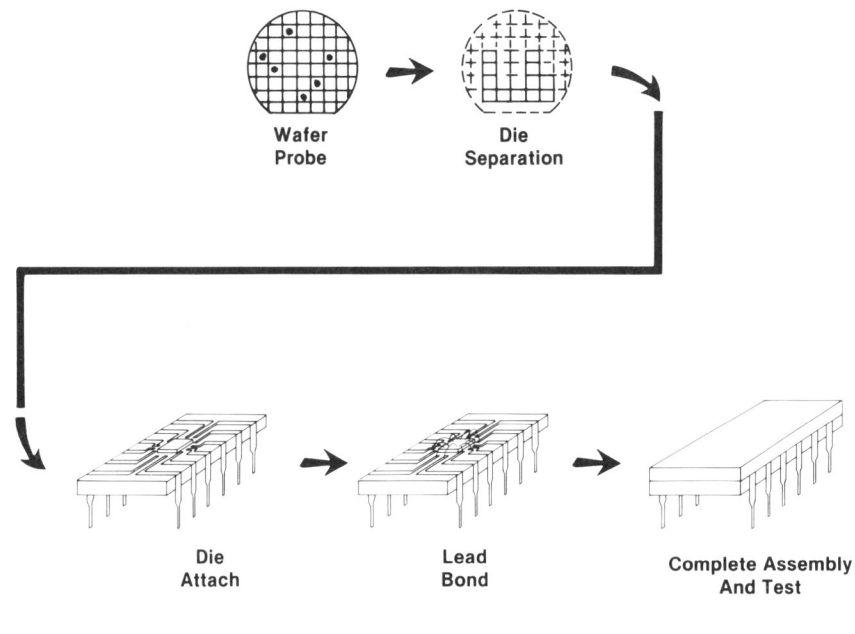

Figure 2-92. IC Assembly Sequence

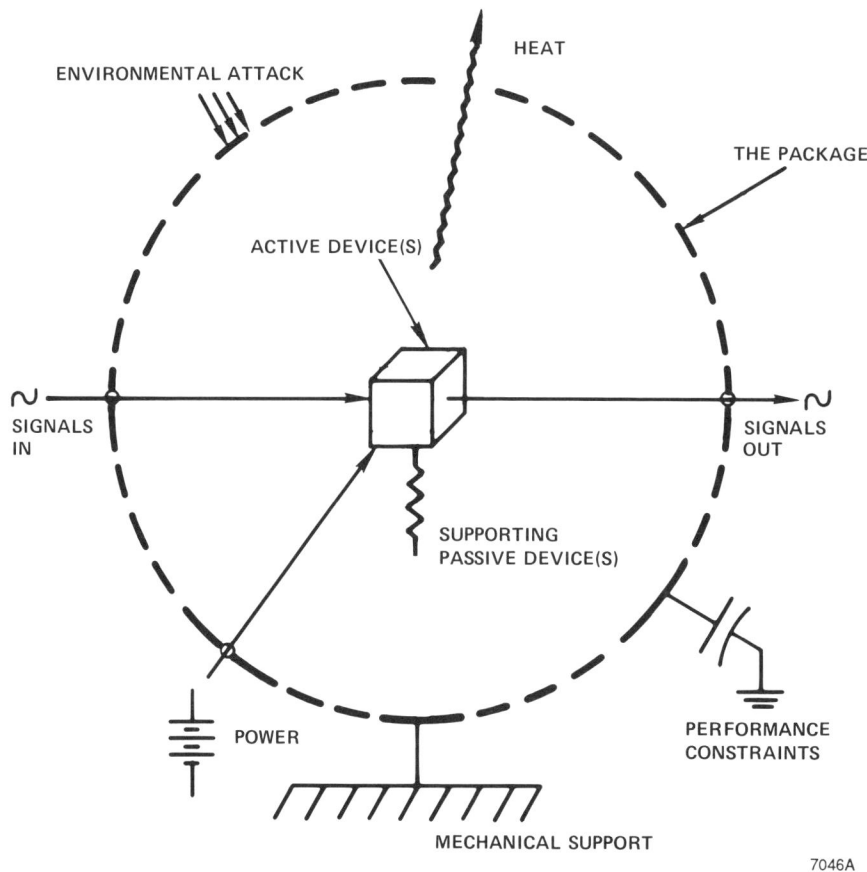

Figure 2-93. The Basis for Electronic Packaging

Basic Integrated Circuit Manufacturing

FUNCTIONS:
- INTERCONNECTION
- PHYSICAL SUPPORT
- ENVIRONMENTAL PROTECTION
- HEAT DISSIPATION

CONSTRAINTS:
- PERFORMANCE
- SIZE
- WEIGHT
- TESTABILITY
- RELIABILITY
- COST

Figure 2-94. The Electronic Package

PERFORMANCE
- SIGNAL RISE TIME
- PROPAGATION DELAY
- SYSTEM IMPEDANCE
- ALLOWABLE LRC
- SWITCHING TRANSIENTS

DENSITY
- CHIP SIZE(S)
- SIGNAL I/O's
- GEOMETRY CONSTRAINTS EACH LEVEL
  — BOND PAD SIZE AND PITCH
  — PACKAGE PAD SIZE AND PITCH
  — SUBSTRATE/BOARD
    — LINE WIDTH AND PITCH
    — NUMBER OF LAYERS

MATERIALS
- PACKAGE (Al_2O_3, BeO, METAL, PLASTIC)
- SUBSTRATE/BOARD
- THERMAL EXPANSION MATCH BETWEEN LAYERS

ASSEMBLY
- DIE BOND
- FIRST LEVEL INTERCONNECT
- LID SEAL/ENCAPSULATE
- PACKAGE ATTACH
  — SURFACE/THROUGH HOLE/OTHER
- LEAD ATTACH
- CAVITY UP/DOWN

COST
- PACKAGE MATERIALS
- CAPITAL COSTS
- LABOR
- YIELD

Figure 2-95. Package Design Factors

Basic Integrated Circuit Manufacturing

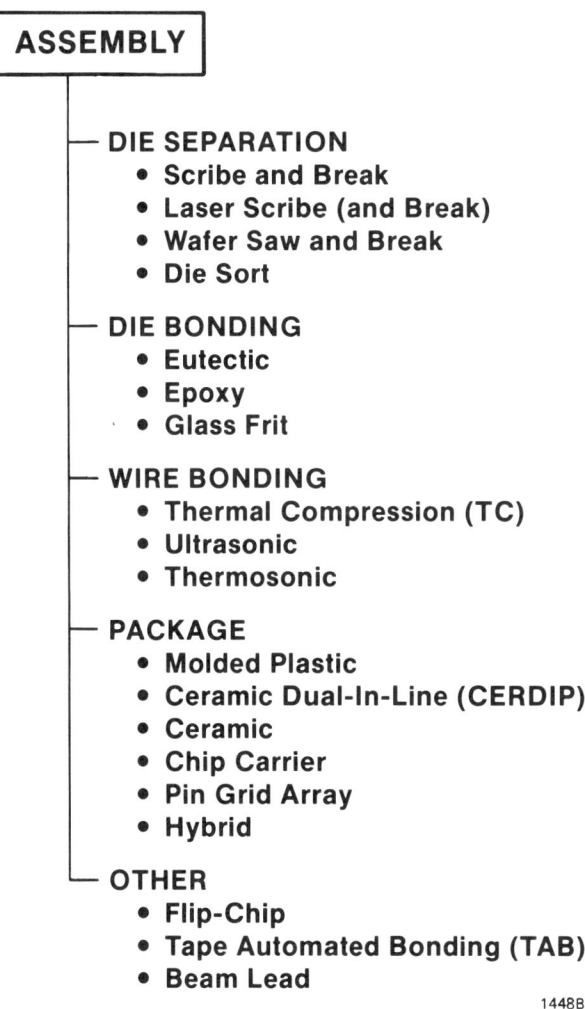

Figure 2-96. Assembly Processes

1. Die Separation

Sawing is the most preferred die separation technique. Cassettes of wafers can be placed into the automatic film mounter wherein each wafer will be properly oriented and placed onto an adhesive-type film that is secured to a rigid frame. This frame can be either metal or plastic. The mounted wafer is transferred to another cassette for transporting to saw.

The handling system in the saw transfers the frame onto the saw chuck. A pattern recognition system orients the frame properly and the correct sawing parameters are selected. The parameters have been previously loaded into the saw's computer memory. The saw will saw the wafer completely thru the wafer thickness in both directions, clean the slurry from the wafer surface and transfer the frame into an outgoing cassette. The dice are held in place by the adhesive on the tape. Visual inspection of the sawing operation is performed per the details of the SPC

Basic Integrated Circuit Manufacturing

The wafer saw is illustrated in Figure 2-97. An example of dice after sawing is shown in Figure 2-98. If the wafers are not sawn completely through their thickness, then a breaking process is used to separate the dice. This is done by applying pressure to the backside of the wafer.

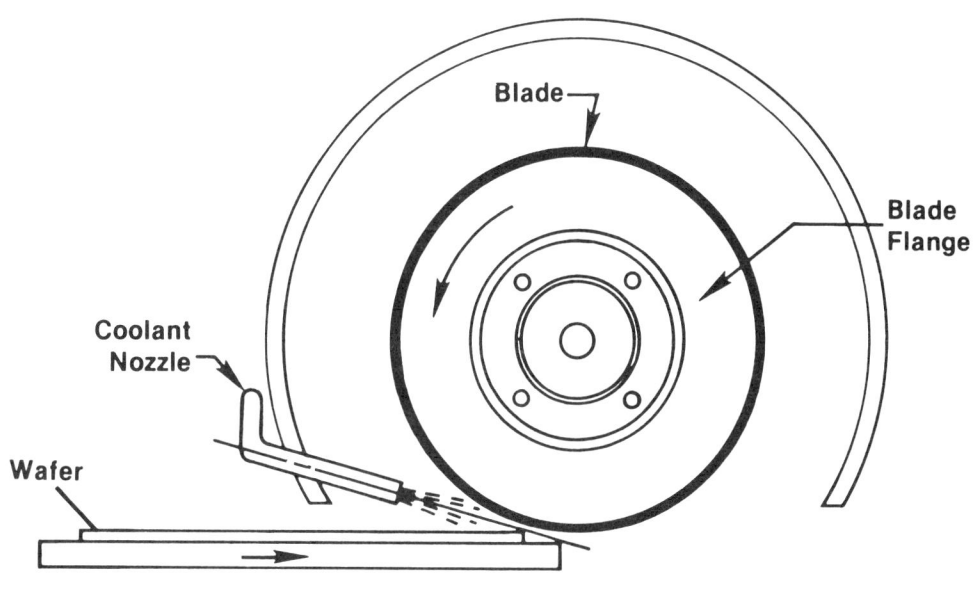

Figure 2-97. Wafer Saw

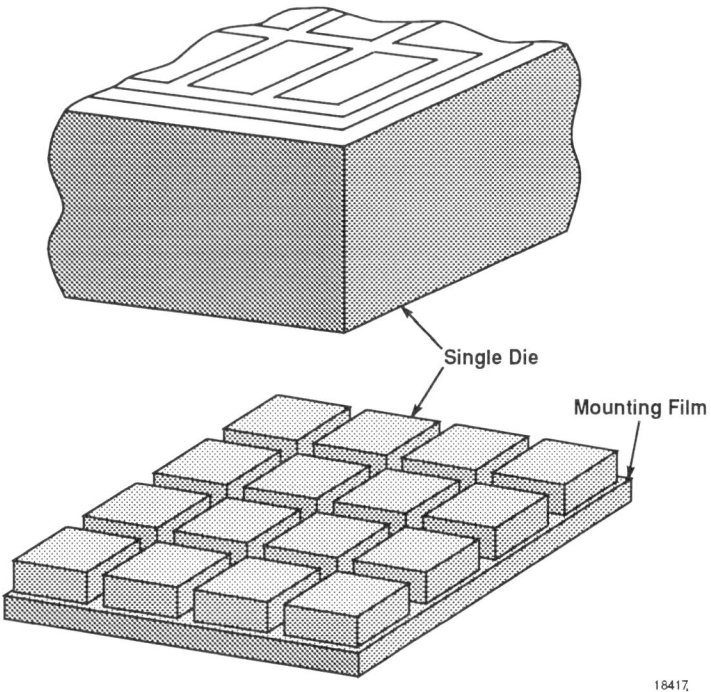

Figure 2-98. Examples of Wafer Saw Cut

2. Die Attach (Bond)

The sawn wafers are transferred to the die attach operation, where the dice are picked up from the adhesive film one at a time and attached to the package substrate or leadframe.

The automated, high-speed die bonder picks up the dice with a collet assembly connected to a vacuum line. The system handles the die only on the edges. This prevents damaging the top surface. The collet assembly transfers the die to the package substrate or leadframe. This method is shown in Figure 2-99.

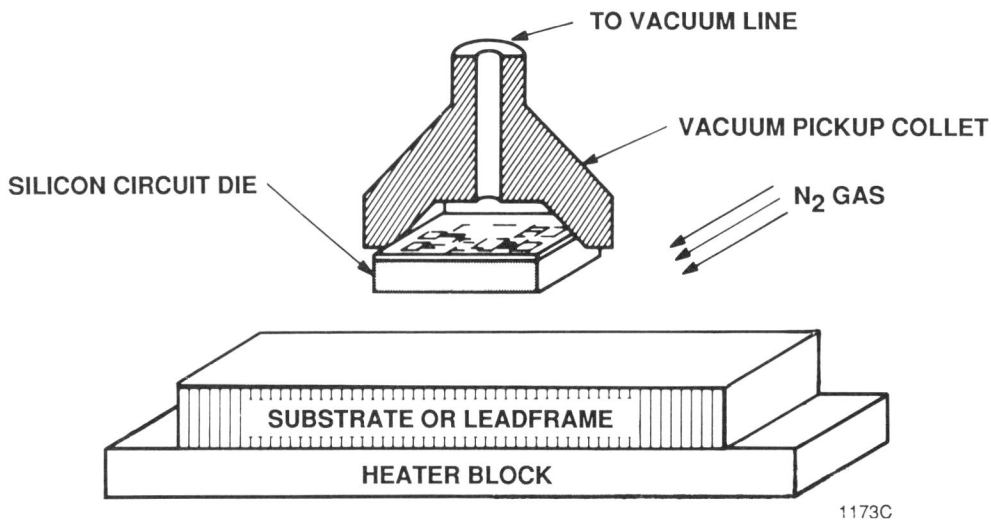

Figure 2-99. Die Bonding

If the die bonder is not equipped with pattern recognition capability and an operator is orienting the die, the operator selects un-inked dice and manually orients the bonder so it places them in the correct position on the package substrate or leadframe.

If the die bonder is equipped with pattern recognition, it automatically picks up only un-inked dice and orients them correctly before placing them on the package substrate or leadframe.

a. Eutectic Attach

There are several different materials and methods used for die attach. For a eutectic attach a thin film of gold is deposited on the back side of the wafer at the end of the wafer fab process. This gold layer is alloyed into the silicon to form a gold-silicon eutectic. The gold-backed die is attached to a package substrate or leadframe having either a gold- or silver-plated area. This attach method is common in ICs needing a good thermal path, mechanical strength, and electrical contact. Many bipolar ICs use this system.

b. Epoxy Attach

The most common attach material is epoxy. It can be formulated with or without silver. If the attach only requires a mechanical connection with modest thermal demands, then a straight epoxy is used. The epoxy will be silver-filled if the attach requires either an electrical connection or a lower impedance thermal path. The majority of MOS products use straight epoxy. An appropriate heat cycle cures the epoxy.

c. Glass Frit Attach

The third alternative is to die attach with glass frit material. This material is generally used for die attach in ceramic packages that require a high-temperature hermetic seal. The glass frit system generally uses a silver-filled mixture. A die attach technology comparison is tabulated in Figure 2-100.

DIE ATTACH TECHNIQUE	CONDUCTIVITY	DIE ATTACH OR CURING TEMPERATURE	MAXIMUM POST ATTACH TEMPERATURE	SHELF LIFE	COMMENTS
Si - Au EUTECTIC	• THERMALLY AND ELECTRICALLY CONDUCTIVE	>377° C	≈475° C	INDEFINITE	• VERY EXPENSIVE • REQUIRES DIE SCRUBBING • POTENTIAL FOR VOIDS ON LARGE DIE SIZES • ESTABLISHED PROCESS
EPOXY	• THERMALLY CONDUCTIVE ONLY • THERMALLY AND ELECTRICALLY CONDUCTIVE (SILVER OR GOLD)t (BEARING)	120° C TO 175° C	300° C TO 400° C	8 MONTHS TO 12 MONTHS ----------- POT LIFE FOR TWO COMPONENT MATERIALS: 4 HOURS TO 4 DAYS	• INEXPENSIVE • MAY REQUIRE TWO COMPONENT SYSTEM (RESIN AND HARDENER) • NO SCRUBBING • ESTABLISHED PROCESS
GLASS FRIT	• THERMALLY AND ELECTRICALLY CONDUCTIVE (SILVER BEARING)	70° C TO 120° C DRYING FOLLOWED BY 420° C TO 450° C FUSION	≈500° C	9 MONTHS	• MODERATELY EXPENSIVE • ONE COMPONENT SYSTEM • NO SCRUBBING • RELATIVELY NEW PROCESS FOR SILVER-BEARING GLASS

Figure 2-100. Die Attach Technology Comparison

3. Wire Bond

Wire bonding is used to connect the bonding pads (usually aluminum) on the die to the posts (bonding areas) on the package substrate or leadframe (Figure 2-101). The wire is attached to the pad on the die by a bonding tool (needle, capillary) and is fed through the tool as the tool moves from the pad to the post. The tool attaches the wire to the post and severs the wire before repeating the cycle. There are three types of wire bonding: *thermocompresssion, ultrasonic,* and *thermosonic ball bonding.*

Basic Integrated Circuit Manufacturing

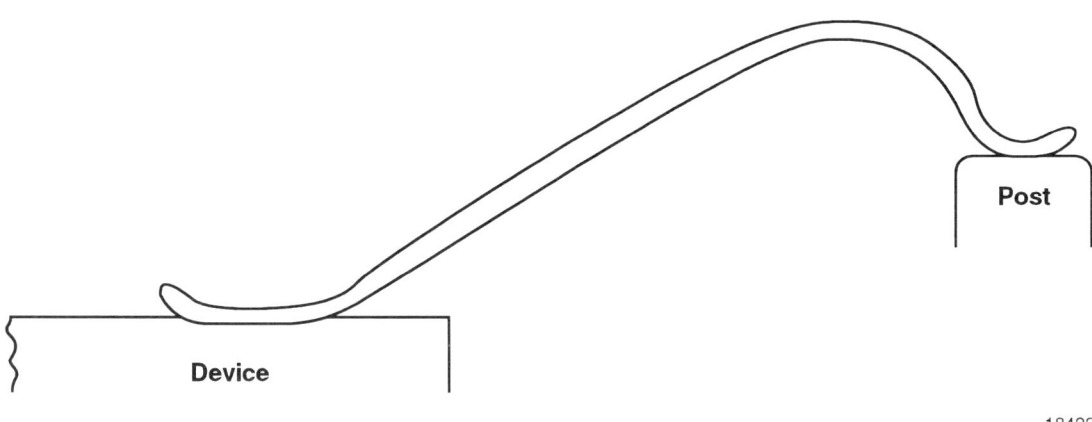

Figure 2-101. Graphic Representation of a Thermocompression Wire Bond

a. Thermocompression

In thermocompression bonding both the package substrate or leadframe and bonding capillary are heated. Pure gold or nearly pure gold wire is used.

The bonding mechanism positions the wire/capillary assembly over the aluminum bonding pad on the die. A force is applied to the wire through the capillary. The combination of the applied force and the heat causes the gold/aluminum interface to blend together to form the bond. The capillary feeds out wire as it moves to the post, where a second bond is made in a similar manner. The thermocompression wedge bond at the bonding pad is shown in Figure 2-102. A graphic representation of the complete wire bond was illustrated in Figure 2-101.

b. Ultrasonic

Some packages require a single-metal system at the die. Since the majority of the ICs produced use aluminum or aluminum alloys for the bonding pads, aluminum wire is needed for wire bonding. This need has been satisfied through the use of ultrasonic energy for the bonding process.

Ultrasonic bonding of aluminum wire is accomplished by feeding the wire through the heel of the bonding capillary (similar to what is used for thermocompression bonding). The wire resides in a groove on the bottom of the capillary. The capillary/wire is positioned over the bonding pad. The capillary applies a force on the wire and ultrasonic energy is transmitted through the capillary to the wire/bonding-pad interface to effect a metallurgical bond. The substrate is not heated. The capillary on the bonding pad is shown in Figure 2-103. The ultrasonic bonding sequence is illustrated in Figure 2-104.

Basic Integrated Circuit Manufacturing

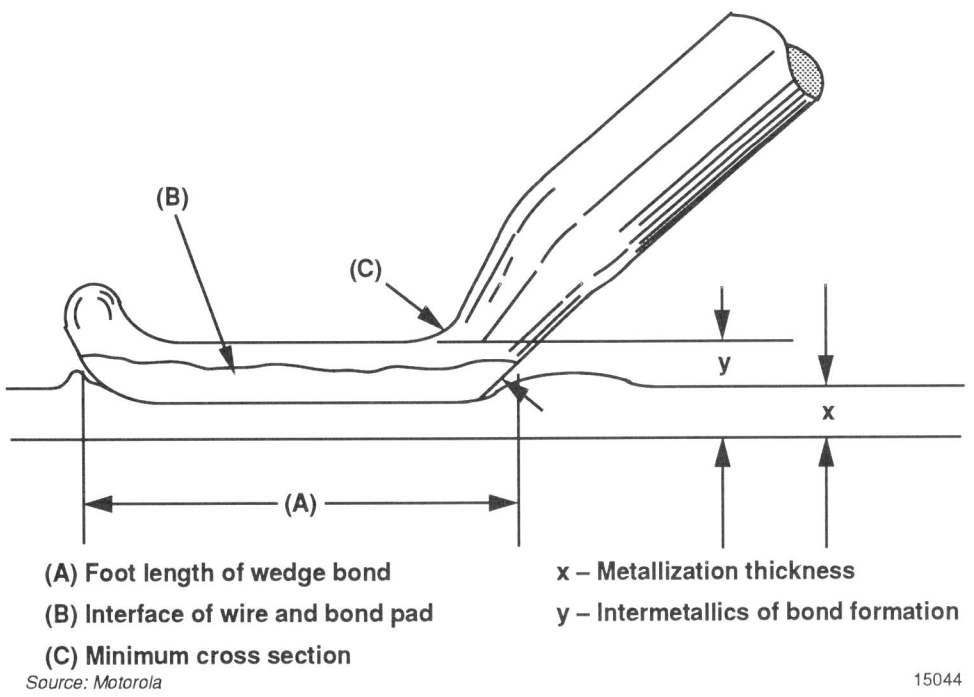

(A) Foot length of wedge bond
(B) Interface of wire and bond pad
(C) Minimum cross section

x – Metallization thickness
y – Intermetallics of bond formation

Source: Motorola

Figure 2-102. Cross Section of Wedge Bond

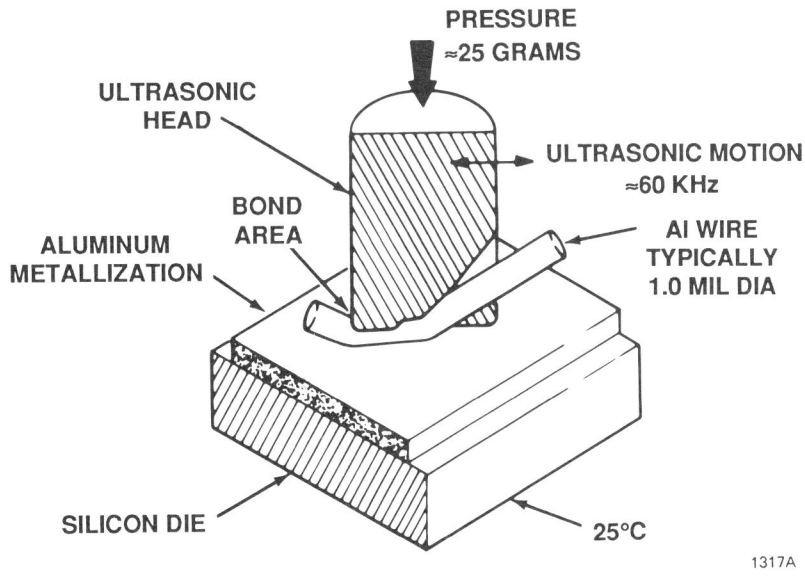

Figure 2-103. Ultrasonic Wire Bonding

Basic Integrated Circuit Manufacturing

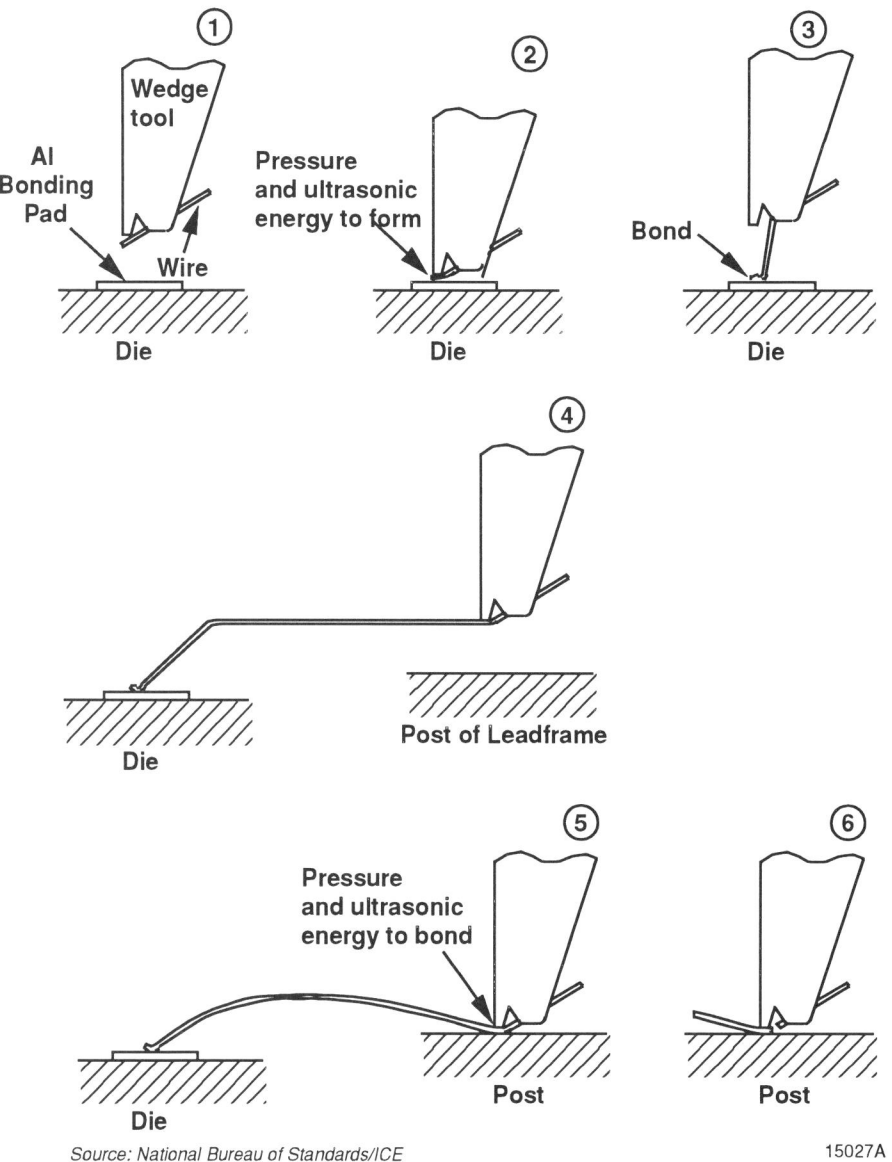

Figure 2-104. Ultrasonic Bonding Sequence

c. Thermosonic Ball Bond

The third type of wire bonding system is called thermosonic ball bonding. Unlike thermocompression and ultrasonic bonding capillaries, the ball-bond capillary has the gold wire fed vertically through a center bore from the top exiting out the bottom. This is illustrated in Figure 2-105. At the beginning of a bond cycle, the wire sticks out from the capillary. It is heated and the melted wire forms a ball (Figure 2-106, No. 2). The ball is positioned on the bonding pad. Pressure is applied to the capillary and ultrasonic energy is transmitted through the capillary. The package substrate or leadframe is heated. The substrate heat is transmitted to the die. The combination of heat, pressure and ultrasonic energy creates a metallurgical bond between the gold bonding wire and the aluminum metallization. The capillary is moved to the package bonding post and properly positioned. The wire is now exiting from one side of the tip of the capillary. Pressure and heat cause the wire to be attached to the post. The second bond shape is partially controlled by the physical shape of the inside bore of the capillary. The inside chamfer (radius) causes the wire to be thinned enough during the second bond such that a slight upward motion at the end of the bonding cycle breaks the wire. The inside chamfer is shown in Figure 2-107. A completed ball bond is shown in Figure 2-108.

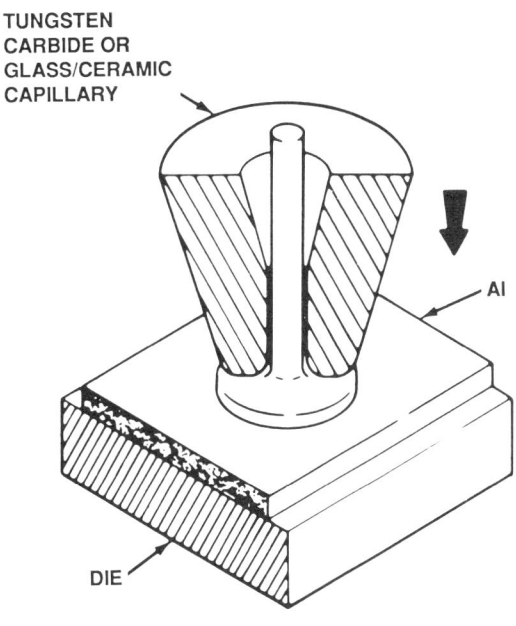

WIRE: 0.0007 IN. TO 0.002 IN. Au
SUBSTRATE: 175 - 225°C
NEEDLE: COLD
GAS: N_2 AT 2 CFH
WEIGHT: 50 GRAMS

1827C

Figure 2-105. Ball Bonding

The ball bonding system has excellent control of the wire loop. Loop control is an important parameter in packaging. The packaging trend continues toward thinner package profiles. The thinner the package the greater the demand on loop control. An example of loop control is illustrated in Figure 2-109.

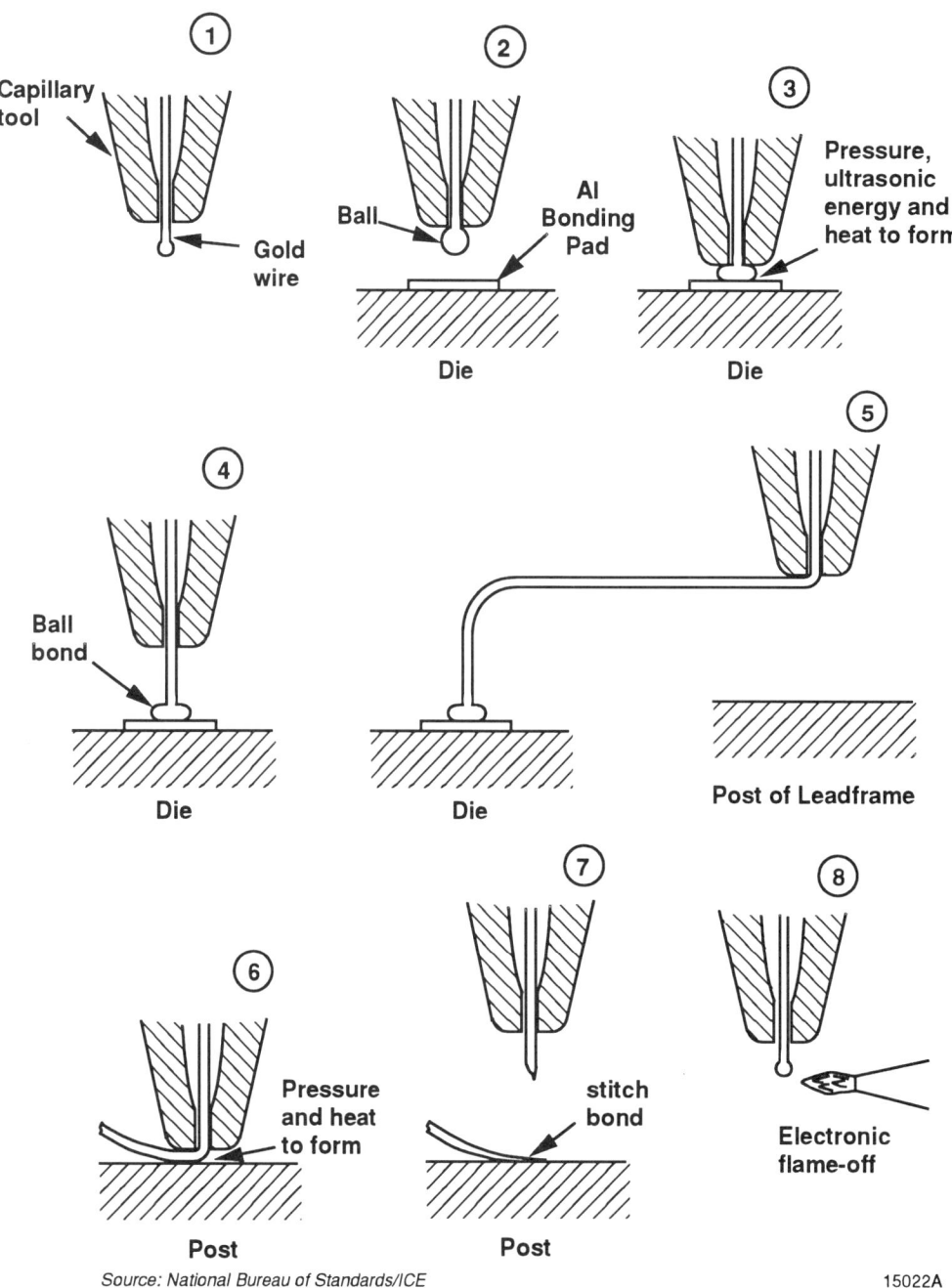

Figure 2-106. Thermocompression/Thermosonic Bonding Sequence

Basic Integrated Circuit Manufacturing

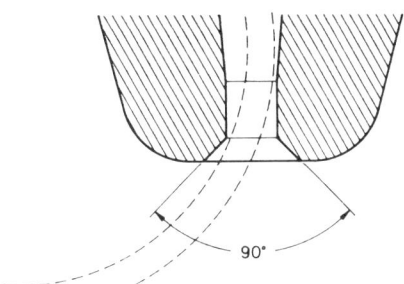

Standard 90° chamfer provides a less sharp transition than the 120° style. This transition causes less drag on the wire allowing improved looping.

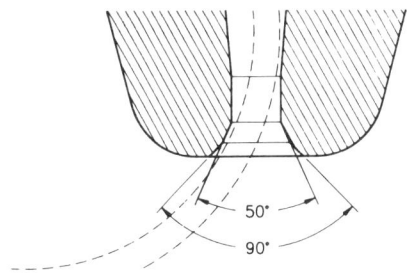

The 90° double chamfer is standard on all Gaiser 90° tools.

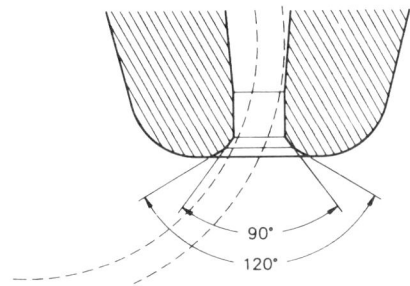

The 120° double chamfer provides a less sharp transition allowing improved looping while producing a strong tail bond.

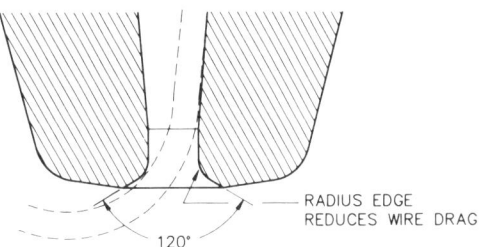

The radiused 120° chamfer provides the best looping while providing the strong tail bond.

Source: Gaiser Tool Company

15067

Figure 2-107. Inside Chamfer Discussion

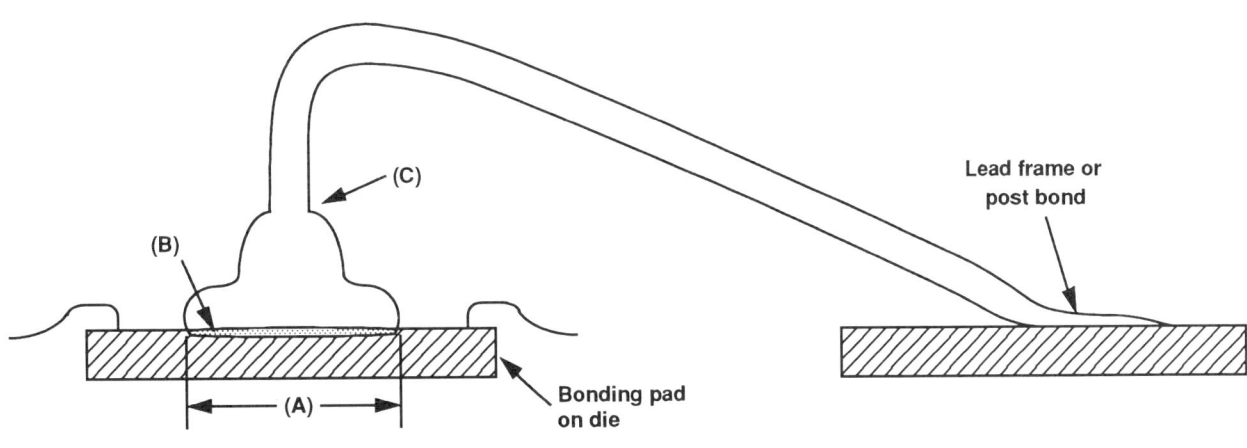

(A) Final ball bond diameter
(B) Interface of wire and bond pad
(C) Neck area of wire

15021A

Figure 2-108. Cross Section of Ball Bond

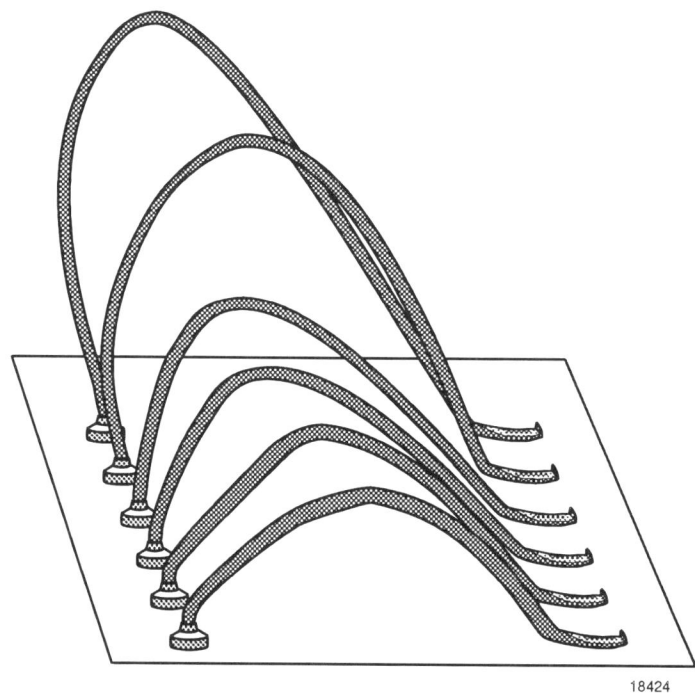

Figure 2-109. Loop Control of Wire Bonds

- Lift at bonding pad metal
- Lift at leadframe
- Break at heel of bond (leadframe)
- Break at ball (neck - down)
- Center wire break

Figure 2-110. Wire Bond Failure Categories

The wire bonding process is controlled by having a good Statistical Process Control program. Part of this program will be a sample visual inspection, wire pull testing, and ball shear testing. The pull test data should be categorized for any bond failure. Generally the five categories shown in Figure 2-110 are used. Each of these categories must be monitored closely, particularly the lifted bond. The ball shear test should serve to guard against lifted bond failures that would not be detected by the conventional wire pull test.

Wire bonding technologies summarized in Figure 2-111. This figure also shows the better productivity of the thermosonic bonding system.

One of the future challenges facing the wire bond method of lead attach is the need to bond to smaller bonding pads placed closer together. Part of the spacing issue centers on the use of staggered bonding pads, often referred to as "zero pitch" (Figure 2-112).

Basic Integrated Circuit Manufacturing

WIRE BOND TECHNIQUE	PROCESS PARAMETERS	WIRE	AUTOMATED ALIGNMENT WITH PATTERN RECOGNITION	WIRE BONDS PER SECOND (TYPICAL)	AUTOMATIC LOOP CONTROL	COMMENTS

ULTRASONIC	- AMBIENT TEMPERATURE
	- ULTRASONIC GENERATOR
	  - 60 Khz
	  - 2 TO 5 WATTS
	- BONDING FORCE
	  20 GRAMS TO 150 GRAMS
		Al OR Al/Si
		1 MIL TO 2 MIL FOR IC
		5 MIL TO 20 MIL FOR POWER
			YES	2 - 3	YES	- ROTARY BONDING HEAD AVAILABLE
						- ALL PARAMETERS FULLY PROGRAMMABLE (X, Y, Z, θ)
						- HIGH RELIABILITY APPLICATION

THERMOCOMPRESSION	- 250°C TO 300°C - BONDING FORCE 20 GRAMS TO 150 GRAMS	Au 0.7 MIL TO 2 MIL	YES	3 - 6	YES	- ALL PARAMETERS FULLY PROGRAMMABLE (X, Y, Z, θ) - ELECTRONIC FLAMEOFF - COMMERCIAL APPLICATIONS - CAN PROMOTE METALLIC INTERDIFFUSION IF BOND TEMPERATURE IS TOO HIGH

THERMOSONIC	- 175°C TO 225°C
	- ULTRASONIC GENERATOR
	  - 60 Khz
	  - 0.5 TO 2 WATTS
	- BONDING FORCE
	  20 GRAMS TO 150 GRAMS
		Au
		0.7 MIL TO 2 MIL
			YES	4 - 8	YES	- ALL PARAMETERS FULLY PROGRAMMABLE (X, Y, Z, θ)
						- ELECTRONIC FLAMEOFF
						- COMMERCIAL APPLICATIONS
						- ALUMINUM WIRE THERMOSONIC BALL BONDING IN FIELD EVALUATION

Figure 2-111. Wire Bond Technology Comparison

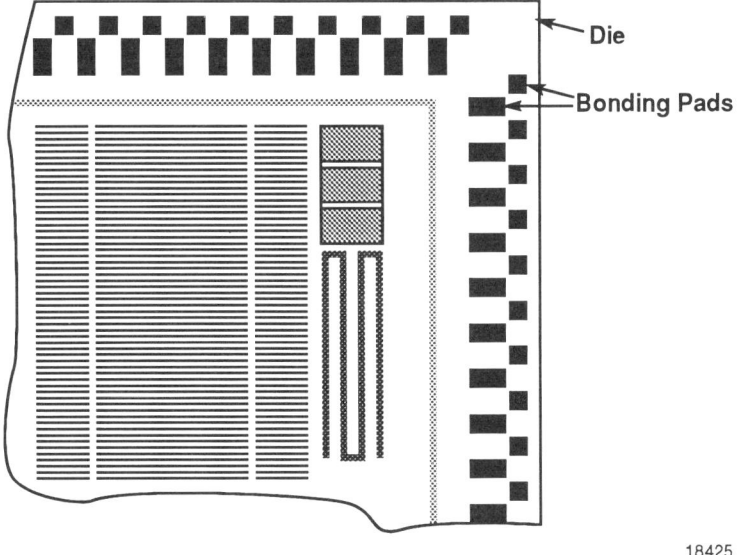

Figure 2-112. Staggered Bonding Pads

4. Plastic Packages

a. Molding

The Plastic Dual-In-Line Package (PDIP) shown in Figure 2-113 has been the industry workhorse since it came into existence in the early 1960's. The PDIP package has served the industry well, rising up to an 80 percent market share in 1981 before beginning to decline in the late 1980's. The assembly process flow chart for the PDIP is shown in Figure 2-114.

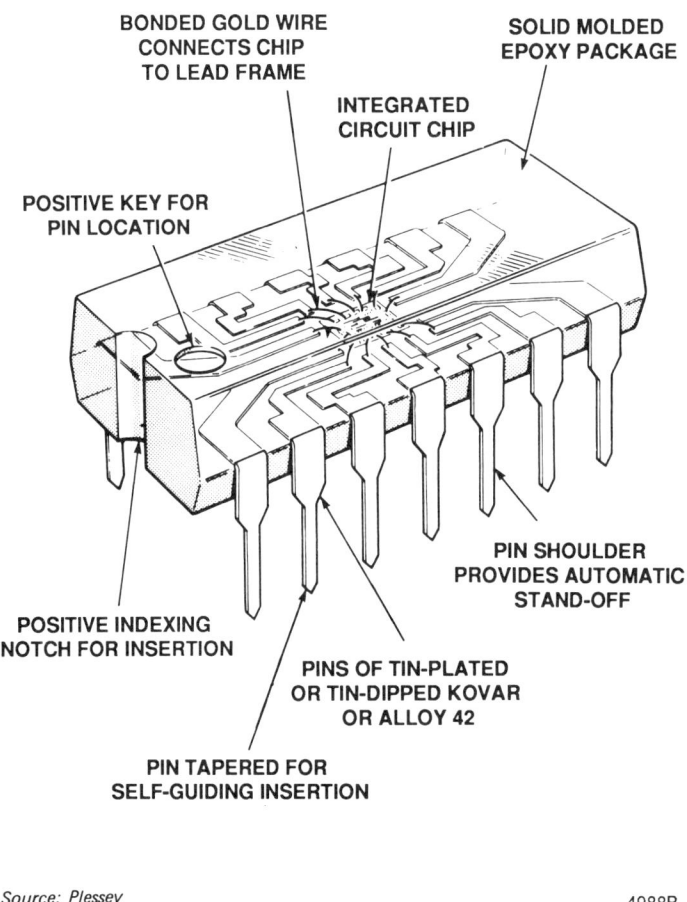

Source: Plessey 4988B

Figure 2-113. Plastic Dual In-Line Package

The plastic assembly method utilizes mass production techniques by using strips of leadframes. This is shown in Figure 2-115. The leadframe strips are loaded into a magazine. The magazine is shown in Figure 2-116. This allows mass handling throughout the assembly process. After die bond and wire bond, the magazine can be interfaced with the various molding process methods. A large block cavity mold is shown in Figure 2-117. The other molding method molds one strip at a time.

Basic Integrated Circuit Manufacturing

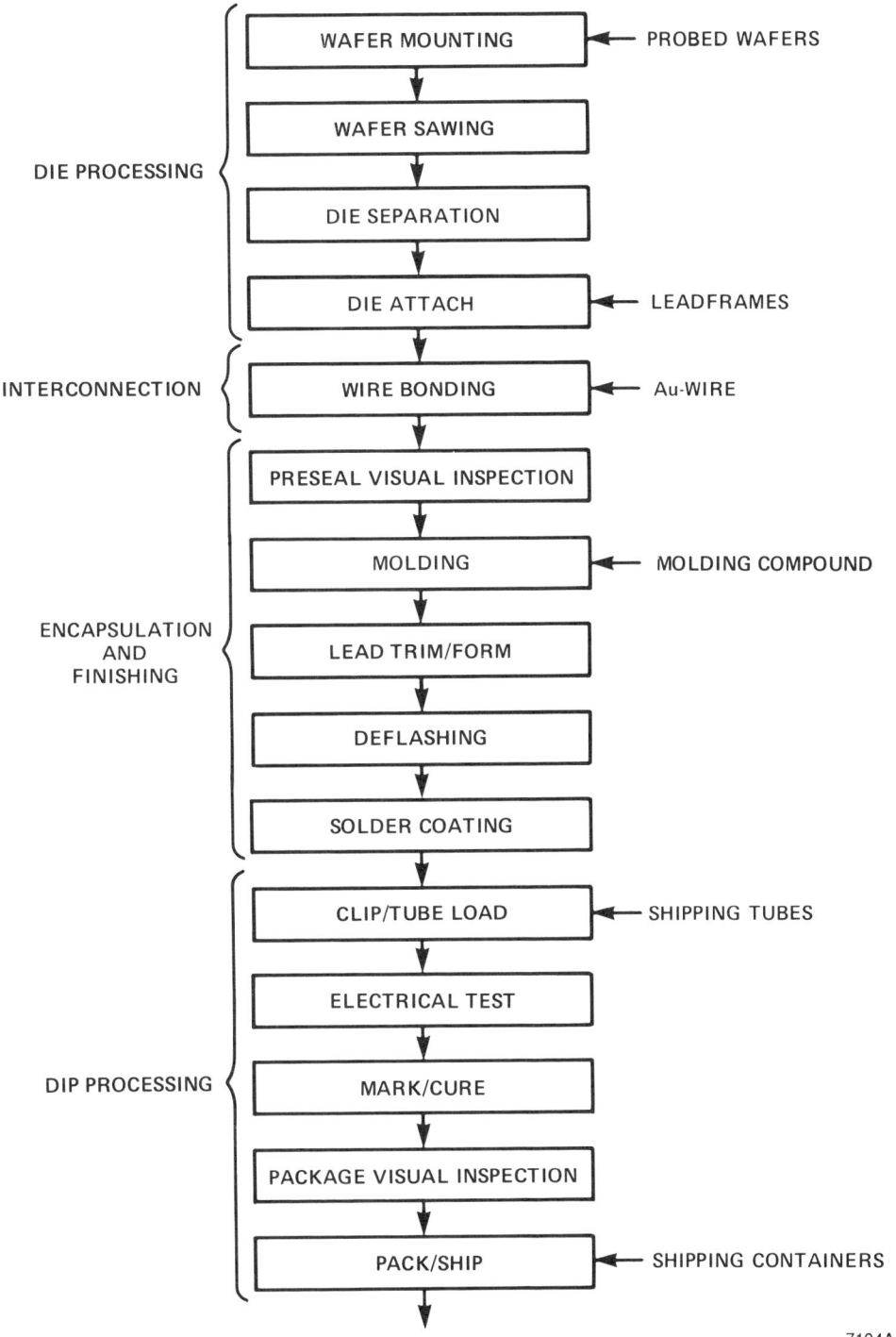

Figure 2-114. Assembly Process Flow Chart for Plastic DIP

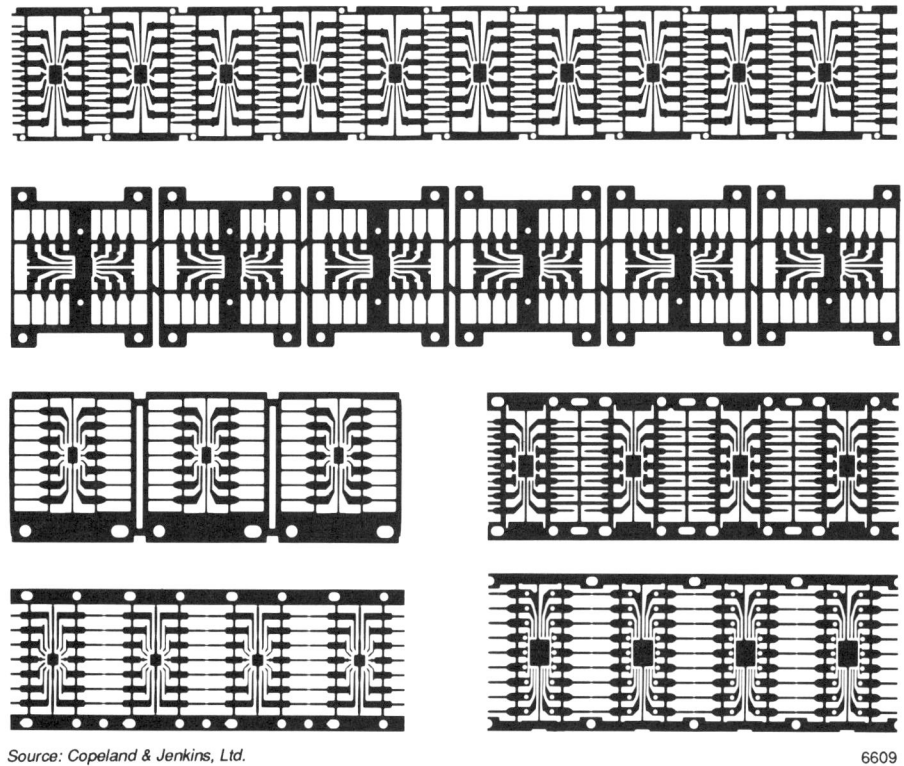

Figure 2-115. Leadframes for Plastic DIP

b. Lead Trim and Form

After the molding operation, the magazines are placed into a large oven to cure the molded material. The strips are transferred to a deflash operation to remove any excess molding materials from the leads of the devices. The next process feeds the strips into the lead trim-and-form tooling. The output from this operation transfers the completed device into a handling tube. Often this handling tube is referred to as a "rail." At this stage the tubes feed a handler that transfers the device onto a "wave soldering" pallet. After the lead soldering operation, the device is returned to the "rail" or tube. The tubes of devices are ready for final package testing.

The previously described processes have been adapted to all of the various plastic package designs, i.e., SOIC, PLCC, PQFP, etc. This has resulted in a cost effective assembly manufacturing.

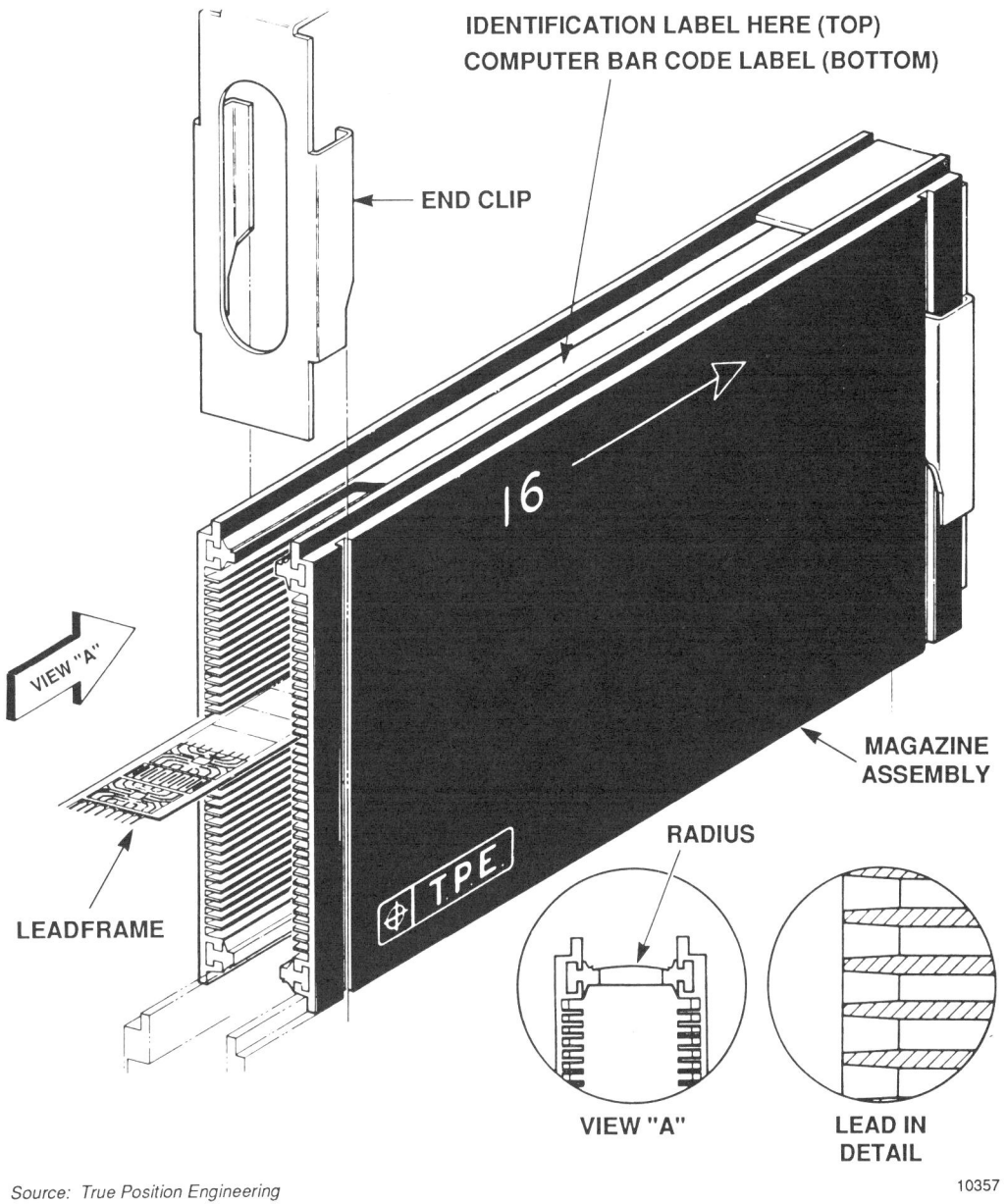

Figure 2-116. Typical Leadframe Magazine

5. CERDIP

The Ceramic Dual-In-Line (CERDIP) package is the most popular ceramic package. An assembly flow chart is shown in Figure 2-118. This is a hermetic package. Many of the handling techniques used in the plastic package assembly have been adopted into ceramic assembly. Because of the hermetic seal, the CERDIP leadframe is lead formed and clipped into individual units after the leadframe is manufactured.

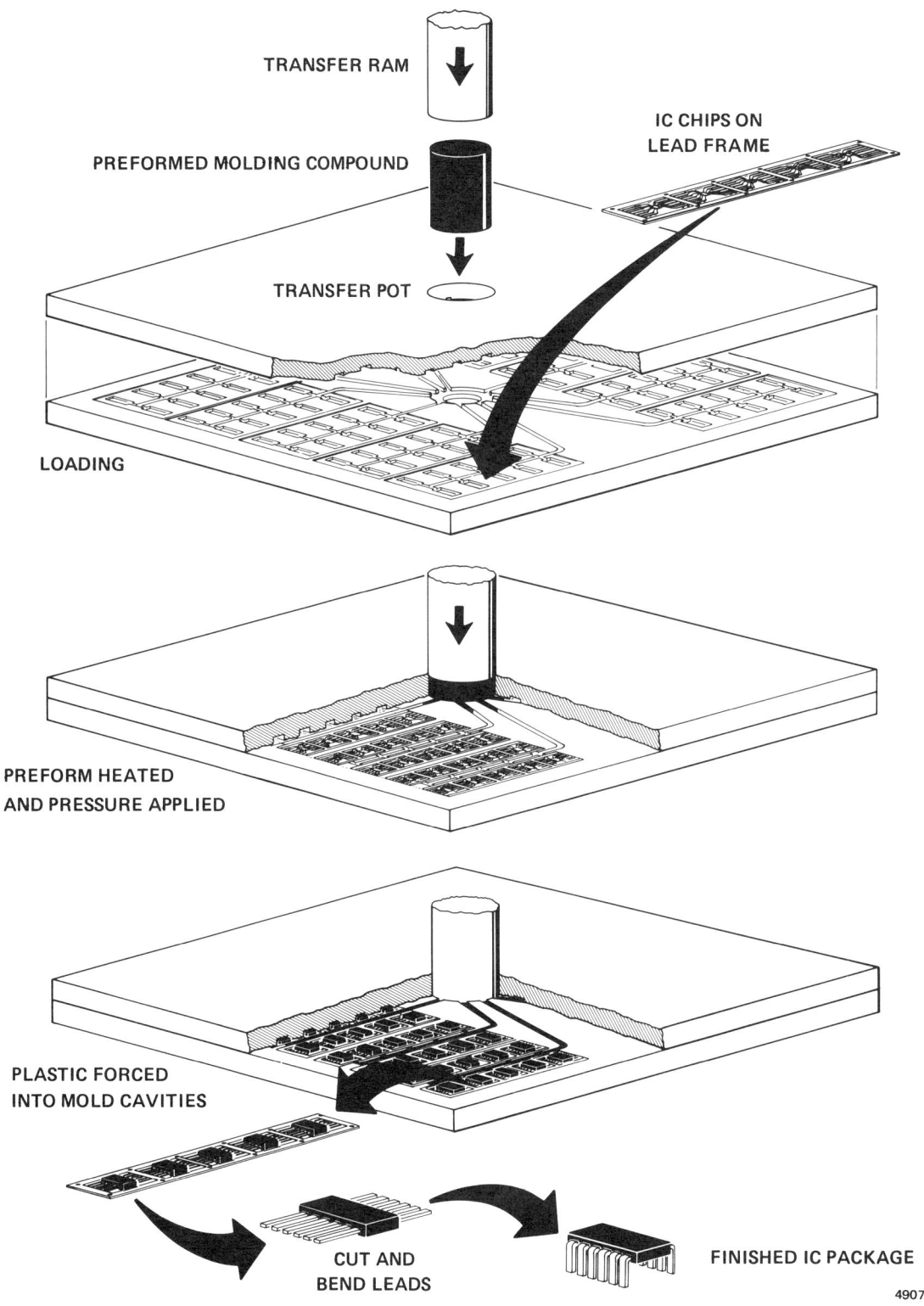

Figure 2-117. Transfer Molding Procedure

Basic Integrated Circuit Manufacturing

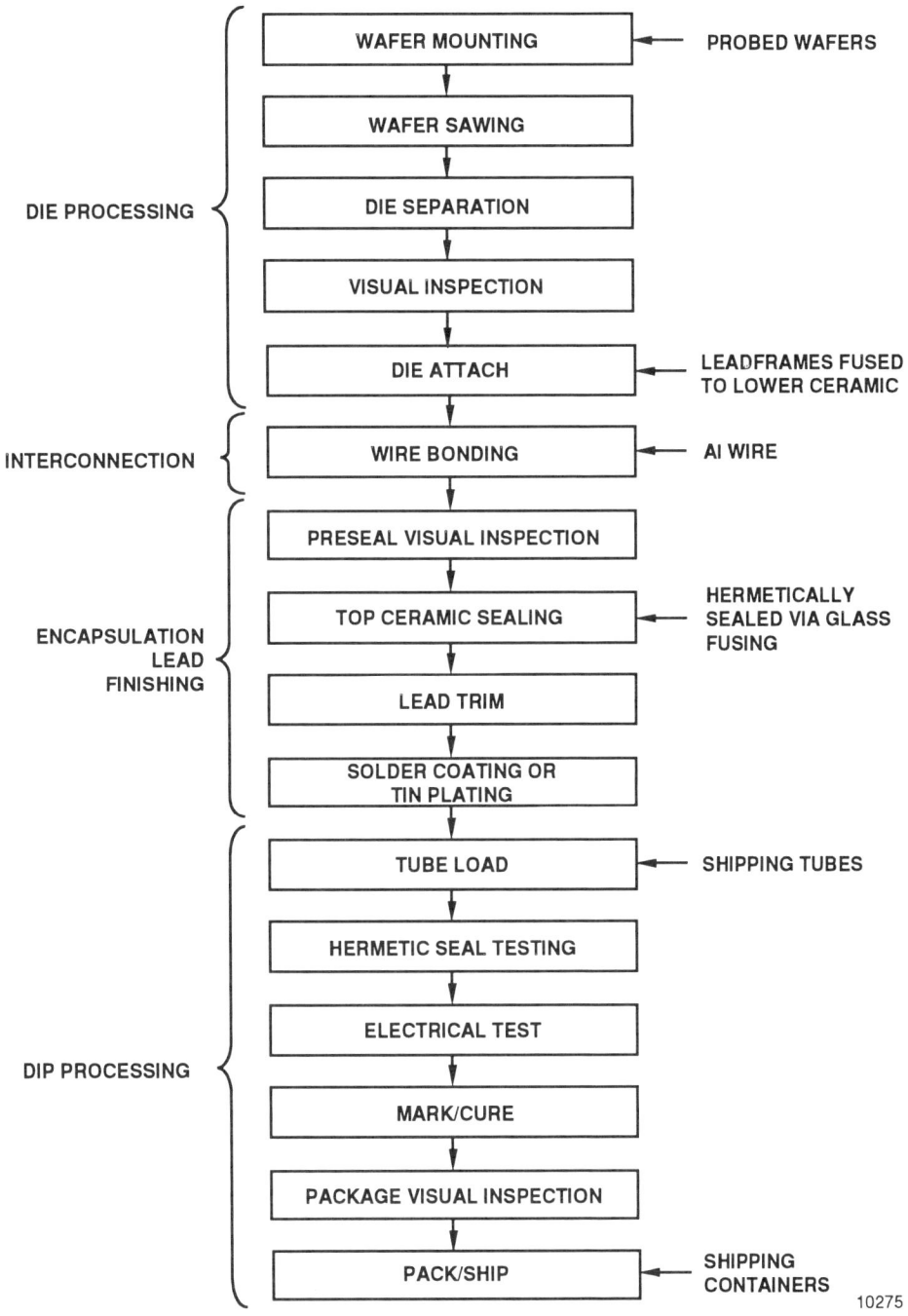

Figure 2-118. Assembly Process Flow Chart for CERDIP

Basic Integrated Circuit Manufacturing

A bottom ceramic with a glass glaze is placed on a heated stage and the leadframe is aligned and embedded into the soft glaze (Figure 2-119, No. 2). The die is oriented and eutecticly attached to the lower ceramic.

The units are then wire bonded and sealed (Figure 2-119, Nos. 2, 3). A glass frit is used to attach the top and bottom of the package together. The melted glass frit provides a hermetic seal when it cools. After the leads of the units are plated, they are trimmed (Figure 2-119, No. 5).

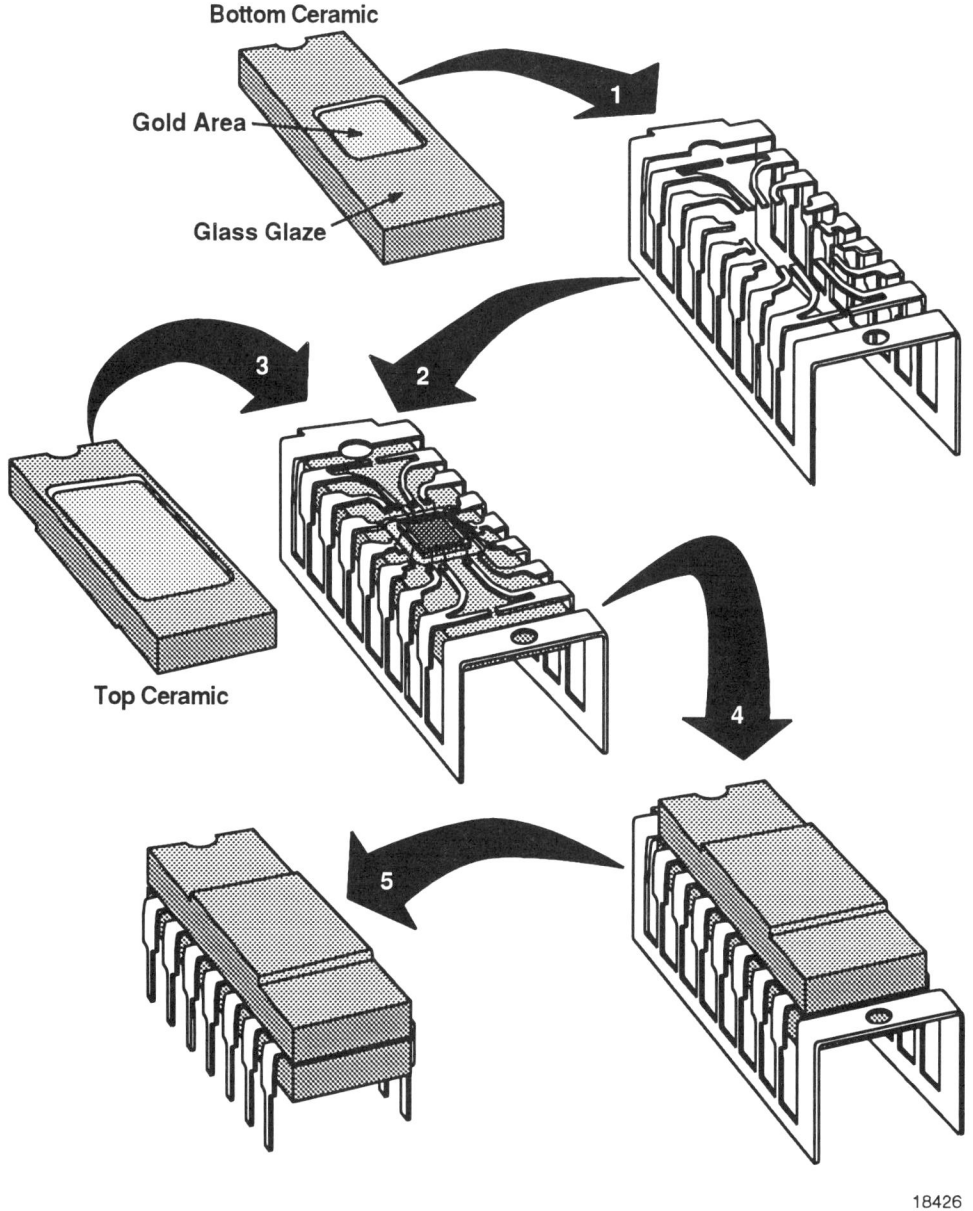

Figure 2-119. CERDIP Assembly Sequence

The final testing of the packaged device is done by automated handlers interfaced with the appropriate tester. In many test areas, the marking equipment is configured in-line with the testing to complete the manufacturing cycle. An in-line Q.C./Q.A. function is integrated into the test area.

There are many other package types used for ICs. The assembly processes for these packages are modifications of the plastic and ceramic assembly methods already presented.

K. PROBE AND TEST

1. Wafer Testing

a. Test Patterns

The electrical evaluation of an IC in wafer form requires two different types of testing. The first electrical test is performed on *"test pattern"* structures. The other electrical evaluation is the testing of the IC to a given electrical specification.

The *"test pattern"* structures are arrangements of individual test patterns, each designed to evaluate specific process attributes. For example, there will be enhancement mode and depletion mode transistors of various geometrical sizes to represent the circuit elements, resistivity structures, critical dimension and overlay registration structures, defect density structures, interconnect test structures for metal step coverage, contact resistance, electromigration, stress migration, and parasitic circuit structures.

All of these test structures' electrical characteristics are measured, recorded, and correlated to the various real-time test wafer measurements taken after each critical process step is completed.

The maturity of the product, the specific technology, the type of photo equipment, and the engineering management will determine where the test structures are arranged on the wafer and the statistics used to accept individual wafers or the wafer lot. Generally the *"test patterns"* are sampled statistically on the wafer and wafers in a lot.

If a sample fails the electrical criteria, the lot will be judged by engineering for further action. Some of the possibilities are:

1. Test additional samples
2. Test all wafers
3. Reject wafers failing the criteria
4. Reject the lot

b. Probe Testing

The wafers receive a 100 percent wafer probe evaluation after the "*test patterns*" are evaluated. This testing is product specific to data sheet or customer requirements. This test of the wafer has been primarily a room temperature test. However, there is a trend to raise the substrate temperature to 50°C. Higher temperatures will cause temperature-sensitive parameters to change, allowing a more critical evaluation.

Each whole die on the wafer is tested. The probe machine moves the wafer to a location under a set of probe needles. The probe needles are lined up with the bonding pads of the first die to be tested. The needles are lowered onto the die (or the die is raised to come into contact with the needles) and that die is tested. The needles separate from the die and the wafer moves to a position where the needles are lined up with another die. The cycle is repeated until all of the whole dice on the wafer are tested. The dice that fail the test are identified, usually with a drop of ink.

After the 100 percent wafer testing is completed, the wafers are sent to the assembly process for final packaging.

2. Final (Product) Testing

The final testing of the product after packaging tests the product to the data sheet specification or to a customer-specific test requirement. A variety of final tests are performed to check for specified electrical characteristics.

Automation is used in a majority of the final test activities. Only when a package is new to the industry or customized and does not fit existing hardware will hand insertion to a socket be used. Bent leads or poor lead finish on a package can cause severe problems for final test. Thus, automation is necessary to prevent these and other problems.

3. Burn-in

"*Burn-in*" typically follows Final Test. The use of "*burn-in*" evolved from semiconductors used in military systems in the early 1960's. The concept has stayed with semiconductor manufacturing in varying degrees. The reason for applying voltage to a semiconductor device at an elevated temperature for some period of time is illustrated in Figure 2-120. This is considered a "*stress test*." From a reliability point of view the elevated temperature with voltage applied for a given time period will cause marginal devices to fail. This removes a potential "*field failure*." The screening is intended to improve the statistical level of reliability by removing the "*infant motality*" devices.

This "*burn-in*" screen is used in many different ways by the semiconductor industry. Within a given company the "*burn-in*" conditions vary over products and technologies.

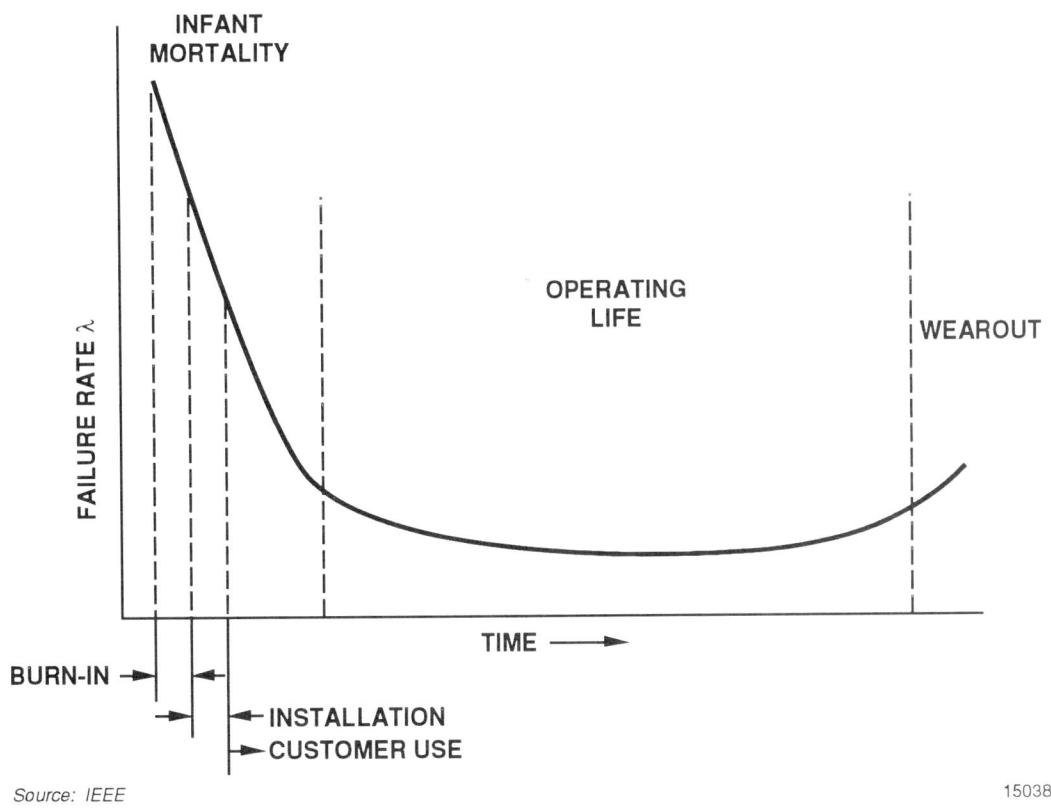

Figure 2-120. Bathtub Curve Prediction of Reliability

L. REVIEW OF SEMICONDUCTOR MANUFACTURING

The impact of integrated circuits on the electronics industry has been phenomenal. Not only are they changing the physical appearance and modes of operation of electronic equipment, but are also causing major changes in the entire structure of the electronic industry. Methods of transacting business are changing. The supplier-user interface is changing. The decision to custom design ICs or use standard or semicustom devices becomes a major consideration for many electronic companies.

The IC manufacturing process roadmap is illustrated in Figure 2-121. The logic designer can either work for a systems company or be part of the IC design team within the semiconductor manufacturer. The circuit designer generally works for the semiconductor manufacturer. The circuit designer translates the logic designer's requirements into a semiconductor circuit design.

The circuit designer will convert the electrical schematic of the circuit into the physical size of each component, i.e., transistor, diode, resistor, capacitor, etc., that makes up the circuit. The designer uses a workstation (CAD) to accomplish the design and do the many different simulations required for design verification.

Basic Integrated Circuit Manufacturing

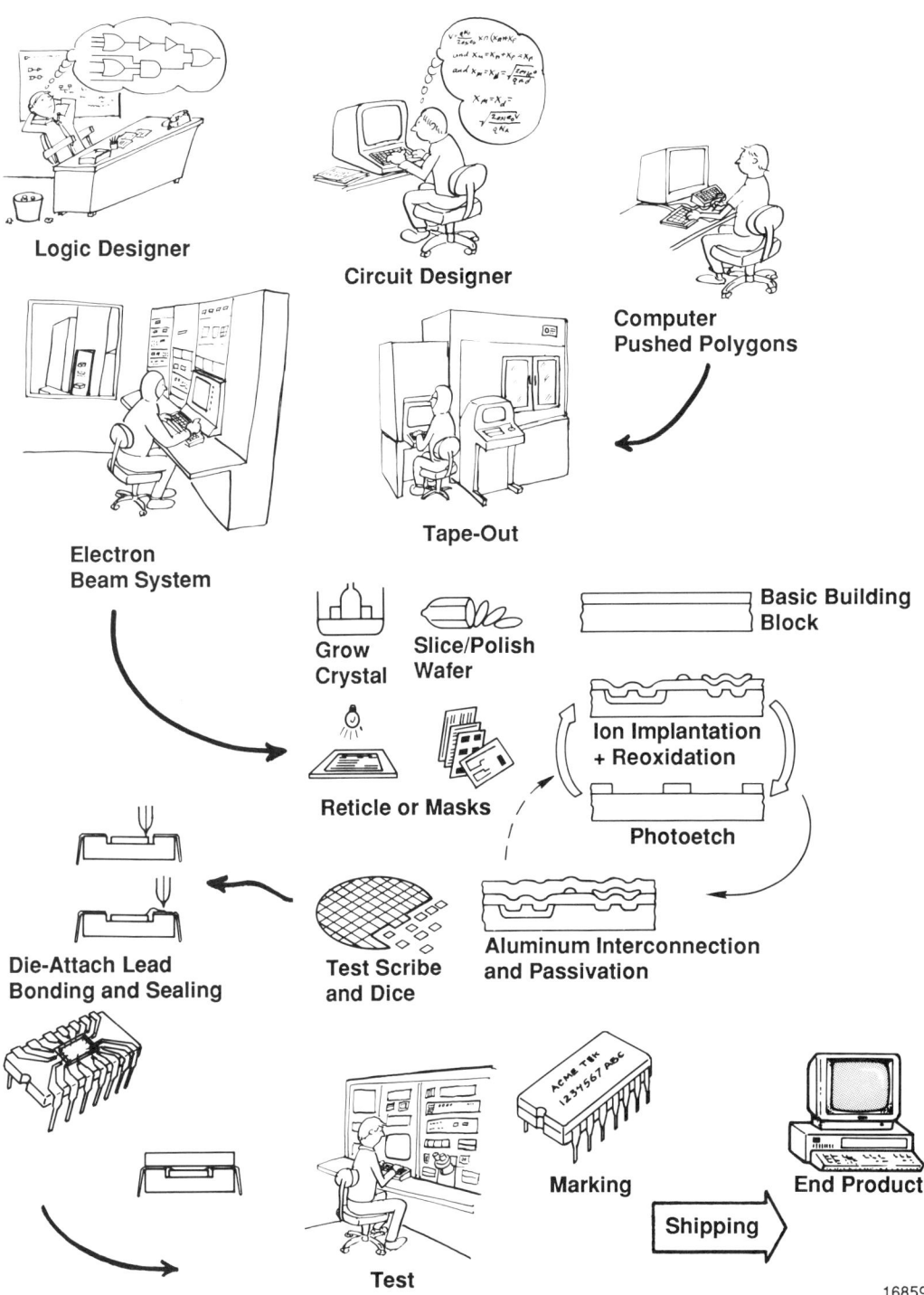

Figure 2-121. Integrated Circuit Manufacturing Process

The geometrical layout is the final output of the workstation in the form of a database tape. The database tape is the input information for the electron-beam system. The electron-beam system uses the input data to create the reticles or masks required in IC manufacturing. The number of reticles or masks is determined by the actual manufacturing process cycle. This is often referred to as the "Fab Process." The newer Fab Processes use between fourteen and twenty-four reticles or masks.

The central region of Figure 2-121 depicts the wafer fab process. After the wafer fab process is completed, the wafers are tested electrically to the required specifications and then forwarded to the assembly process.

The assembly process will package each electrically good die. After packaging is completed, the device will be given a final electrical test, burn-in as required, and shipped to the end user. The end user creates the electronic system by bringing together all the necessary electrical components, mechanical hardware and the final package of the product.

There is no doubt the world is in the silicon age. Silicon devices in the form of discrete products or ICs touch our lives in many ways everyday. And — since silicon is the second most abundant element in the earth, there doesn't appear there will ever be a shortage of the raw material.

3 PACKAGING

Packaging the IC chip is a necessary step in the manufacturing process because the IC chips are small, fragile, susceptible to environmental damage, and too difficult to handle by the IC users. In addition, the package acts as a mechanism to "spread apart" the connections from the tight pitch (center to center spacing of two parallel conductors) on the IC die to the relatively wide pitch required by the Printed Circuit Board (PCB) manufacturer.

A. PAD PITCH

The pad pitch on the IC chip is typically 0.006 inch (6 mils or 152μm). This spacing is already much larger than the 2 to 8 microns (0.08 to 0.31 mils) pitch of the wiring (metallization) on the IC chip. But PCB wiring requires an even larger pitch, usually between 40 and 100 mils. The package acts as a "bridge" between the two sizes, effectively spreading apart the spacing from the IC chip dimensions to the PCB dimensions, as shown in Figure 3-1.

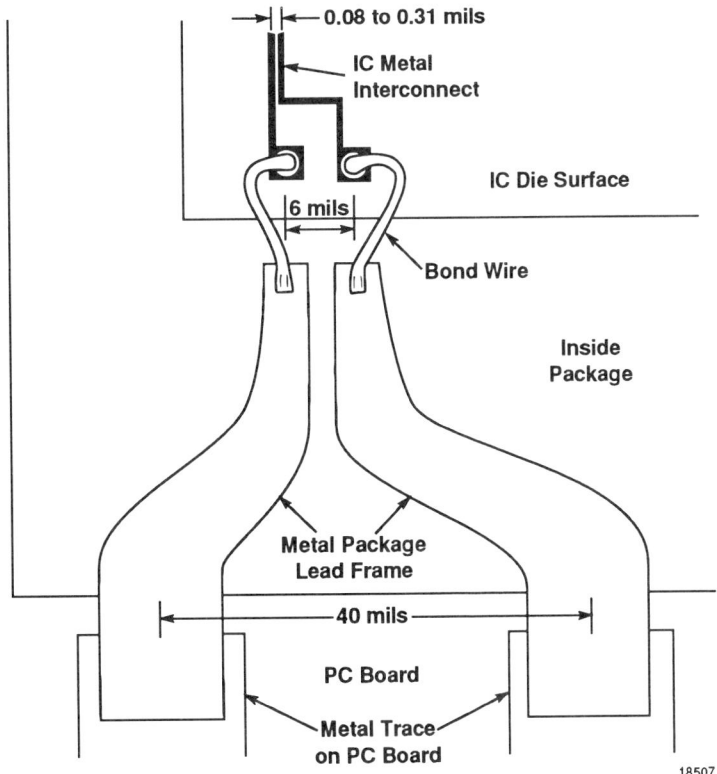

Figure 3-1. IC Chip to PCB Lead Spacing

INTEGRATED CIRCUIT ENGINEERING CORPORATION

Packaging

Early packages were required to expand the pitch to the very large board pitch required for economical manufacture of PCBs used in commercial applications. In addition, the packages were mounted on the PCBs through plated holes, typically on 100 mil pitch. Three package types were used, ceramic and metal for hermetically sealed requirements and plastic for general commercial use. Hermetically sealed parts are ones where the IC chip is enclosed in a sealed compartment and outside chemicals and gases cannot reach the die.

B. DUAL IN-LINE PACKAGES (DIPs)

The case outlines of the plastic and ceramic Dual In-line Packages (DIPs) are nearly identical. The lead configuration consists of two rows of leads, both with 100 mil pitch. The plastic DIP is shown in Figure 3-2. The side-braze and cerdip packages are its hermetic alternatives. The cerdip shown in Figure 3-3 is considerably less expensive than the side-braze shown in Figure 3-4. Both packages are available with quartz windows in the top for EPROM devices so the chip can be erased with ultraviolet light.

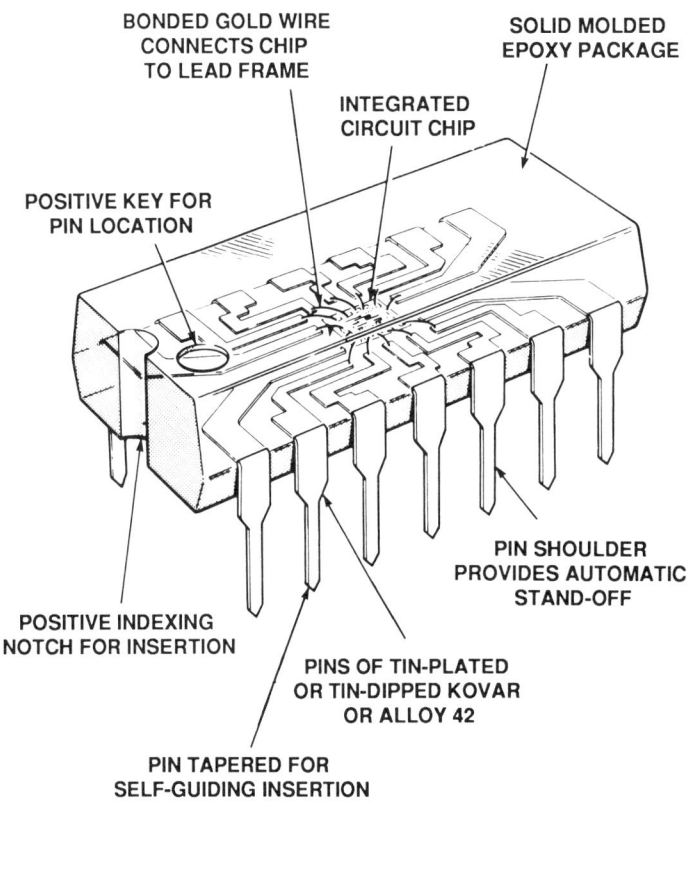

Source: Plessey

Figure 3-2. Plastic Dual In-Line Package

Packaging

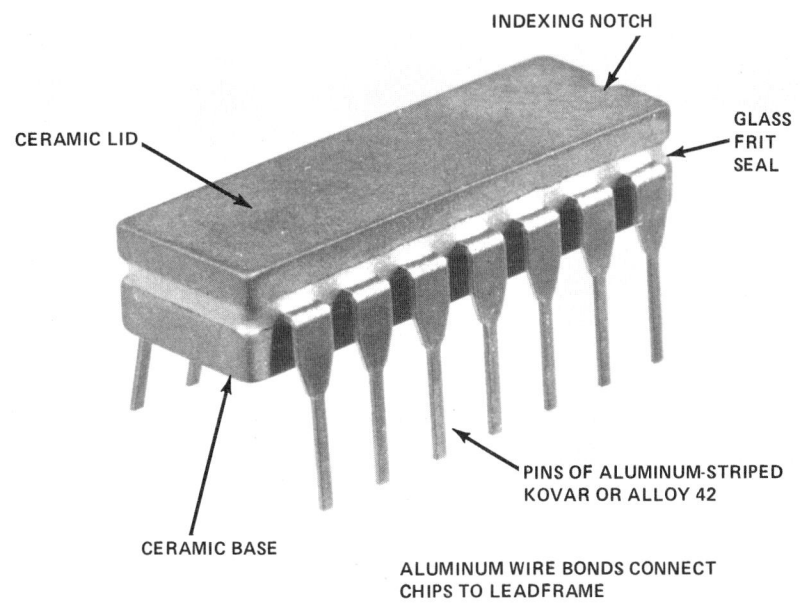

Figure 3-3. Cerdip Package

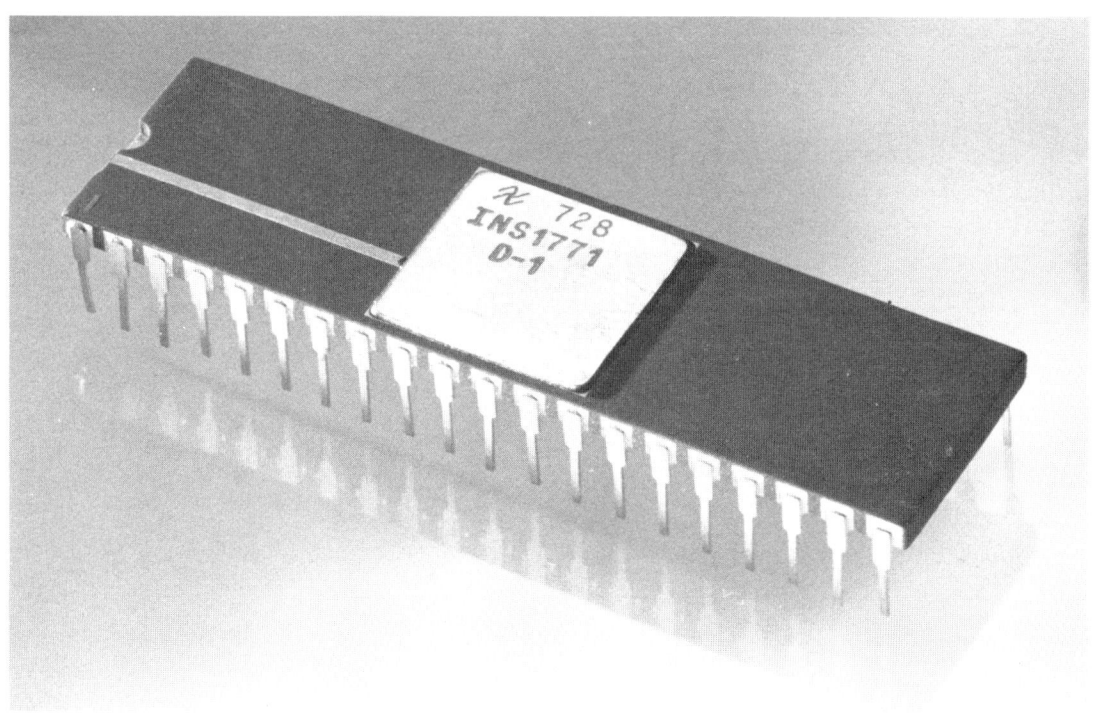

Figure 3-4. Side-Brazed Ceramic DIP

Packaging

C. SINGLE IN-LINE PACKAGES (SIPs)

Another alternative to the DIP is the single in-line package shown in Figure 3-5. To further reduce the PCB cost — which goes up as the wiring pitch is reduced — some manufacturers use the packages with staggered leads, such as the QUIP in Figure 3-6 or the SIP in Figure 3-7.

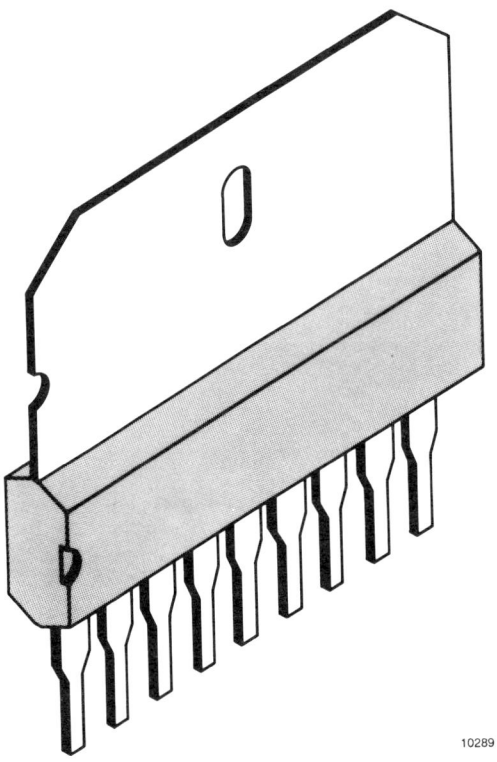

Figure 3-5. 9-Pin Single In-Line Package

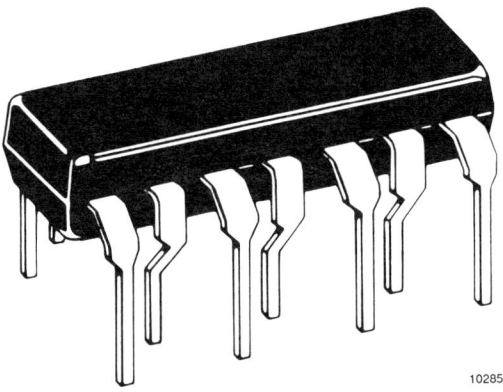

Figure 3-6. 14-Pin Plastic QUIP

Packaging

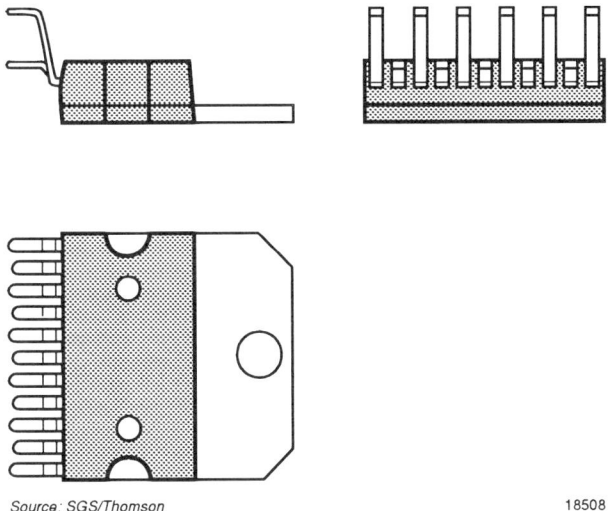

Figure 3-7. Multiwatt 11

D. CERAMIC FLATPACKS

A package used for early ICs in military and space applications was the ceramic flatpack, shown in Figure 3-8. The pitch on this package is 50 mils, allowing smaller system boards. The smaller pitch, however, requires more expensive board processing and more careful IC handling techniques, so the flatpacks were rarely used in commercial applications. These packages are manufactured using the cerdip approach because of the very thin package outline.

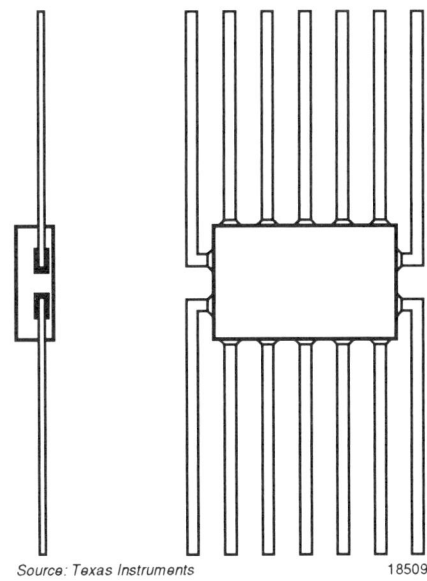

Figure 3-8. Flatpack

E. PIN GRID ARRAYS

The limitation of 100 mil minimum pin spacing for commercial PCBs is a significant problem for two reasons: high pin count IC chip requirements and board space needed for each IC. The first problem is solved by the Pin Grid Array (PGA) shown in Figure 3-9. Here, the pins remain on 100 mil pitch, but cover most of or all of the entire bottom surface of the package. These packages are offered in pin counts from about 100 to 600 pins, and are available with heatsinks (as shown in the figure) to help remove the heat generated inside the package. This can be a problem with fast clock rate microprocessor chips where the frequent switching generates significant heat.

INTEGRATED CIRCUIT ENGINEERING CORPORATION

Packaging

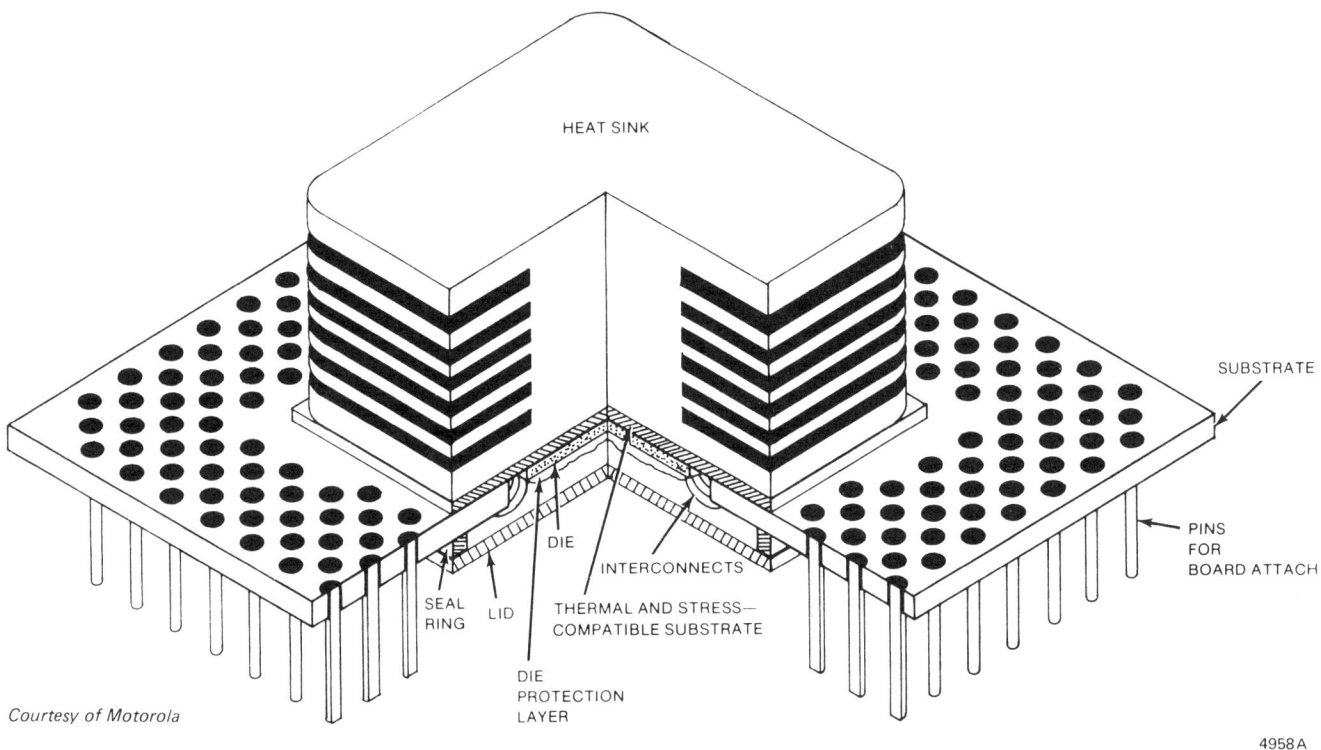

Figure 3-9. Motorola VLSI Package with Die Cavity Below and Finned Heat Sink Above

F. SURFACE MOUNT

1. SOIC, QSOP

The second problem — better use of board space — is solved through the use of surface mount packaging. A comparison of four surface mount packages and one DIP is shown in Figure 3-10. PCB technology improvements have allowed the package pin pitch to be reduced further as shown in the comparison between the Small Outline IC (SOIC) and the Quarter Size Small Outline Package (QSOP) in Figure 3-11.

2. LCC, PLCC, PQFP

Figure 3-12 shows a comparison between the Leadless Chip Carrier (LCC), the Plastic Leaded Chip Carrier (PLCC) and the SOIC. The PLCC and the SOIC use the gull wing lead configuration (sometimes called a Lap Joint). Figure 3-13 shows the most common lead configurations for PLCCs, as well as the advantages and disadvantages of each. Figure 3-14 shows the internal construction of the Plastic Quad Flat Pack (PQFP) with corner bumpers to protect the leads from being damaged.

Packaging

PACKAGE TYPE	DESCRIPTION	ADVANTAGES	DISADVANTAGES
DUAL-IN-LINE	0.100-IN. PIN CENTERS 0.125-IN. PIN LENGTH 0.160-IN. BODY THICKNESS 0.300- TO 0.900-IN. BODY WIDTH	CAPABILITIES GENERALLY EXIST LOWEST IMPLEMENTATION COST MINIMUM ENGINEERING EFFORT WELL-ESTABLISHED RELIABILITY DATABASE	PIN COUNT LIMITED TO LESS THAN 64 I/O LINES NOT SURFACE MOUNTABLE
SOIC (SMALL-OUTLINE IC)	0.050-IN. PIN CENTERS 0.030-IN. PIN LENGTH (SOLDERABLE) 0.098-IN. BODY THICKNESS 0.155-IN. AND 0.300-IN. BODY WIDTH	SURFACE MOUNTABLE LOWEST MATERIAL COST ALLOWS MAGAZINE HANDLING SOLDERING/REWORK (VS QUAD FLAT PACK)	SIGNIFICANT EQUIPMENT DESIGN AND TOOLING HANDLING AND TEST MORE DIFFICULT BOARD ROUTING/SOLDERING/REWORK OF 0.050-IN. PIN CENTERS PELIABILITY QUESTIONS
QUAD FLAT PACK	0.50-IN. PIN CENTERS 0.030-IN. PIN LENGTH 0.095-IN. BODY THICKNESS VARIABLE BODY WIDTH	LOW MATERIALS COST SOLDER JOINTS VISIBLE (VS CHIP CARRIER)	PROCESS AND TEST MODIFICATIONS NEEDED PIN-COUNT LIMITATIONS HANDLING DIFFICULT BOARD ROUTING MORE COMPLEX SOLDERING/REWORK OF 0.050-IN. CENTERS RELIABILITY CONCERNS
CHIP CARRIER	0.050-IN. PIN CENTERS 0.030- AND 0.060-IN. STANDOFF 0.100-IN. BODY THICKNESS JEDEC-COMPATIBLE SIZES	HIGH PIN COUNTS POSSIBLE SURFACE MOUNTABLE LESS STRESSED THAN QUAD FLAT PACK	DEVELOPMENT TIME AND COST BOARD ROUTING/REWORK MORE DIFFICULT SOLDER-CONNECT TECHNOLOGY AND JOINT RELIABILITY
PLASTIC CHIP CARRIER	0.050-IN. PIN CENTERS 0.100-IN. PIN LENGTHS 0.145-IN. BODY THICKNESS JEDEC-COMPATIBLE SIZES	SURFACE MOUNTABLE MINIMUM ENGINEERING EFFORT LOW MATERIALS COST	TECHNOLOGY COMPATIBLE WITH PIN COUNTS UNDER 84 LEADS BOARD ROUTING/REWORK OF 0.050-IN. CENTERS

Source: EDN

Figure 3-10. Package Type Comparisons

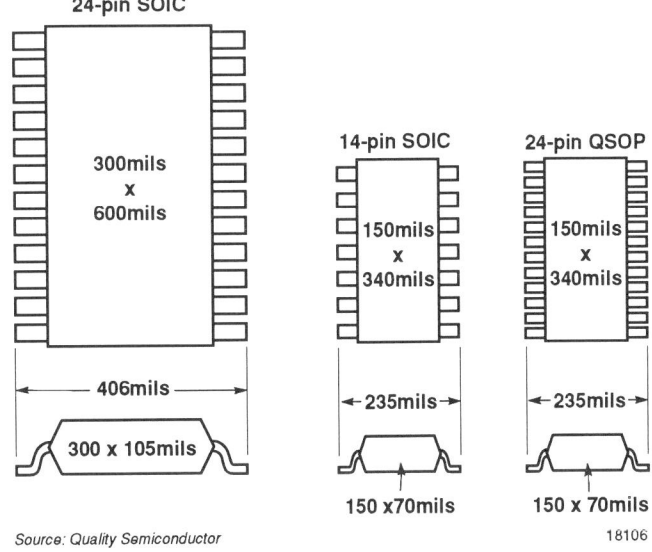

Source: Quality Semiconductor

Figure 3-11. QSOP Versus SOIC Packaging

INTEGRATED CIRCUIT ENGINEERING CORPORATION

Packaging

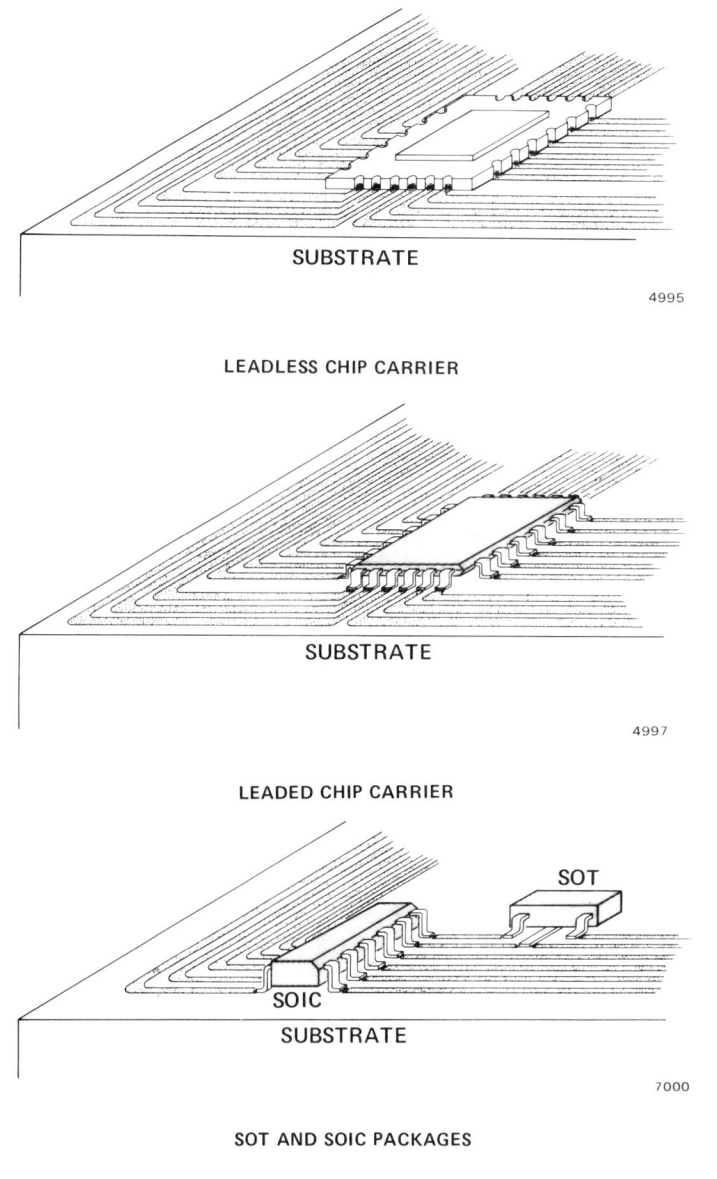

Figure 3-12. Surface-Mount IC Packages

3. Thermal Coefficient of Expansion

One technical compromise is present with all packaging schemes, but becomes very critical with surface mount — mismatch in thermal expansion coefficients of the various elements — the silicon die, the package materials, and the PCB materials. Figure 3-15 shows the expansion coefficients of some of the commonly used materials. As the table shows, most of the materials are inherently incompatible. Therefore, to make the packaging systems work reliably, flexible elements such as package leads are required.

Packaging

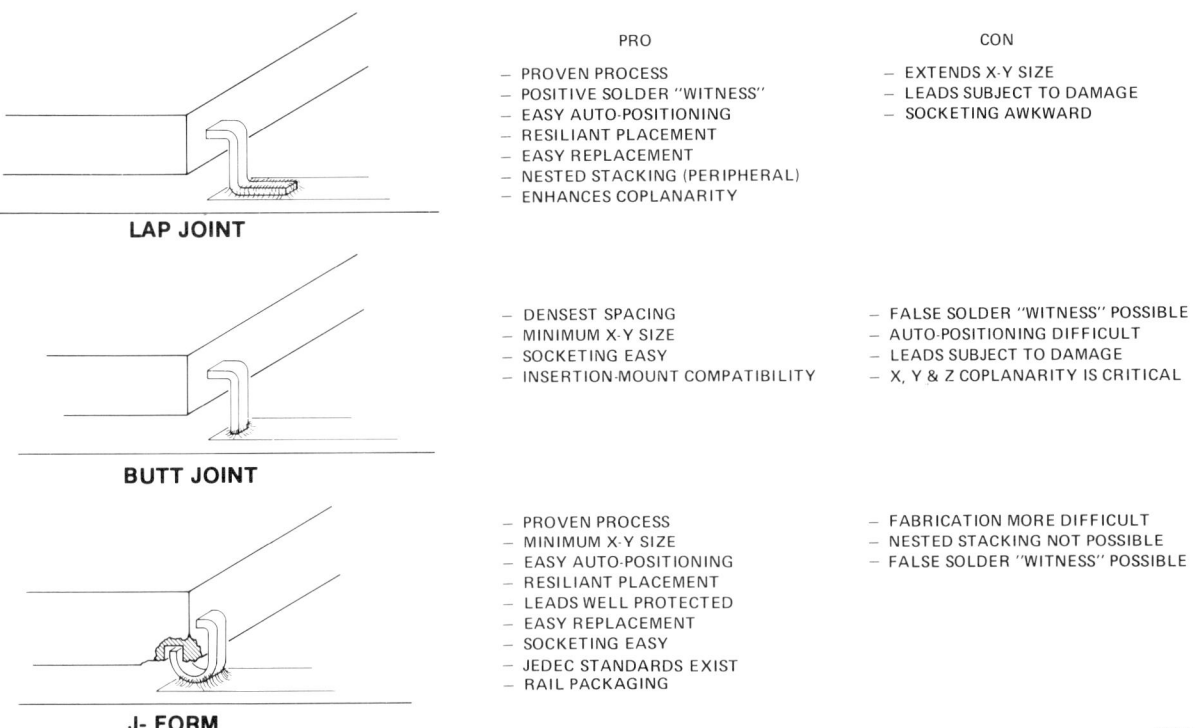

Figure 3-13. Surface Mount Leaded Connections

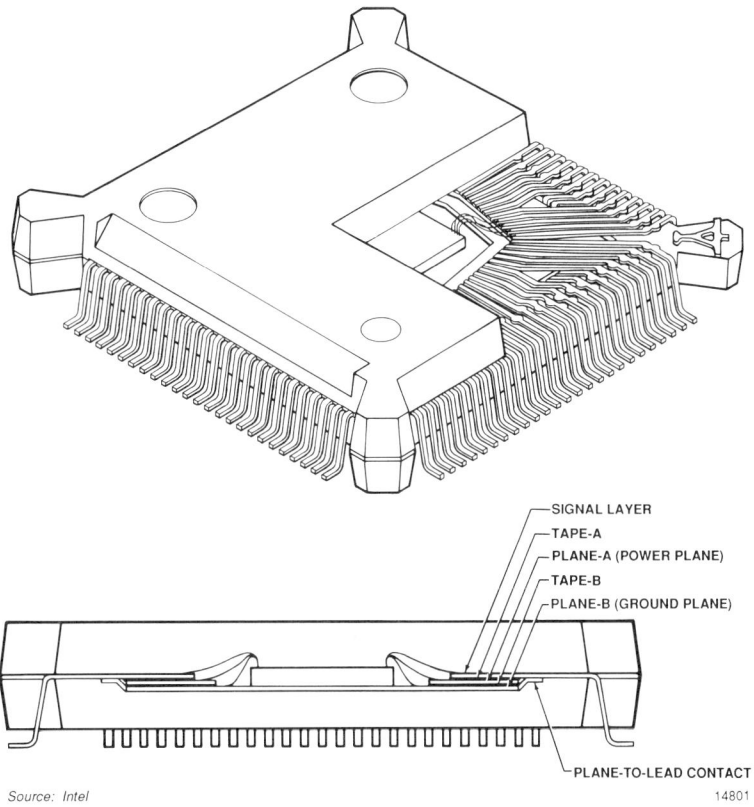

Source: Intel

Figure 3-14. Top and Cross-Section Views of MM/PQFP Package

MATERIAL	DIELECTRIC CONSTANT	THERMAL EXPANSION COEFFICIENT (10^{-7}/°C)	THERMAL CONDUCTIVITY (W/m·°K)	APPROXIMATE PROCESS TEMPERATURE (°C)
NON-ORGANICS				
92% Alumina	9.2	60	18	1,500
96% Alumina	9.4	66	20	1,600
Si_3N_4	7	23	30	1,600
SiC	42	37	270	2,000
AlN	8.8	33	230	1,900
BeO	6.8	68	240	2,000
BN	6.5	37	600	>2,000
Diamond High Pressure	5.7	23	2,000	>2,000
Plasma CVD	3.5	23	400	≈1,000
Glass-Ceramics	4 – 8	30 – 50	5.0	1,000
Copper Clad Invar (10% Copper)/(Glass coated)	—	30	100	800
Glass Coated Steel	6	100	50	1,000
Silicon	11.7	0.3	1.57	1,420
ORGANICS				
Epoxy-Kevlar (x-y) (60%)	3.6	60	0.2	200
Polyimide-Quartz (x-axis)	4.0	118	0.35	200
Fr-4 (x-y plane)	4.7	158	0.2	175
Polyimide	3.5	500	0.2	350
Benzocyclobutene	2.6	350 – 600	0.2	240
Teflon®	2.2	200	0.1	400

Teflon is a trademark of Dupont Company
Source: Microelectronics Packaging – An Overview

Figure 3-15. Properties of Package Insulator Materials

4. Flip Chip, Multi-Chip Modules (MCMs)

If flexible leads are not available, then the package material must be a close match to the substrate material. IBM has used the flip-chip (small solder balls on each bonding pad) approach for over thirty years. It has been successful for mounting the flip chips on ceramic substrates, as shown in Figure 3-16, for single-chip packages, as well as for Multi-Chip Modules (MCMs), as shown in Figure 3-17.

Packaging

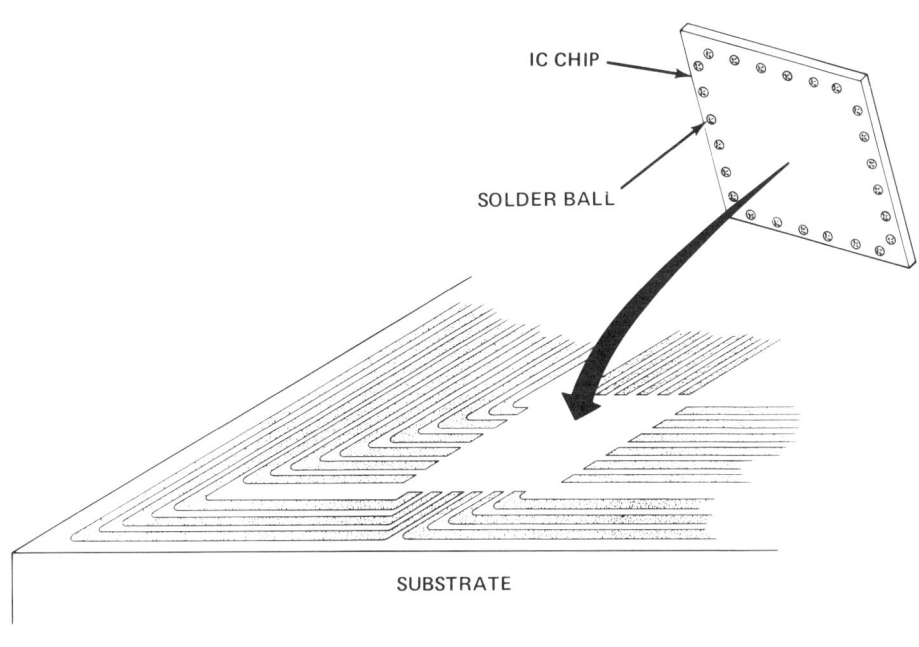

Figure 3-16. Flip-Chip Mounting

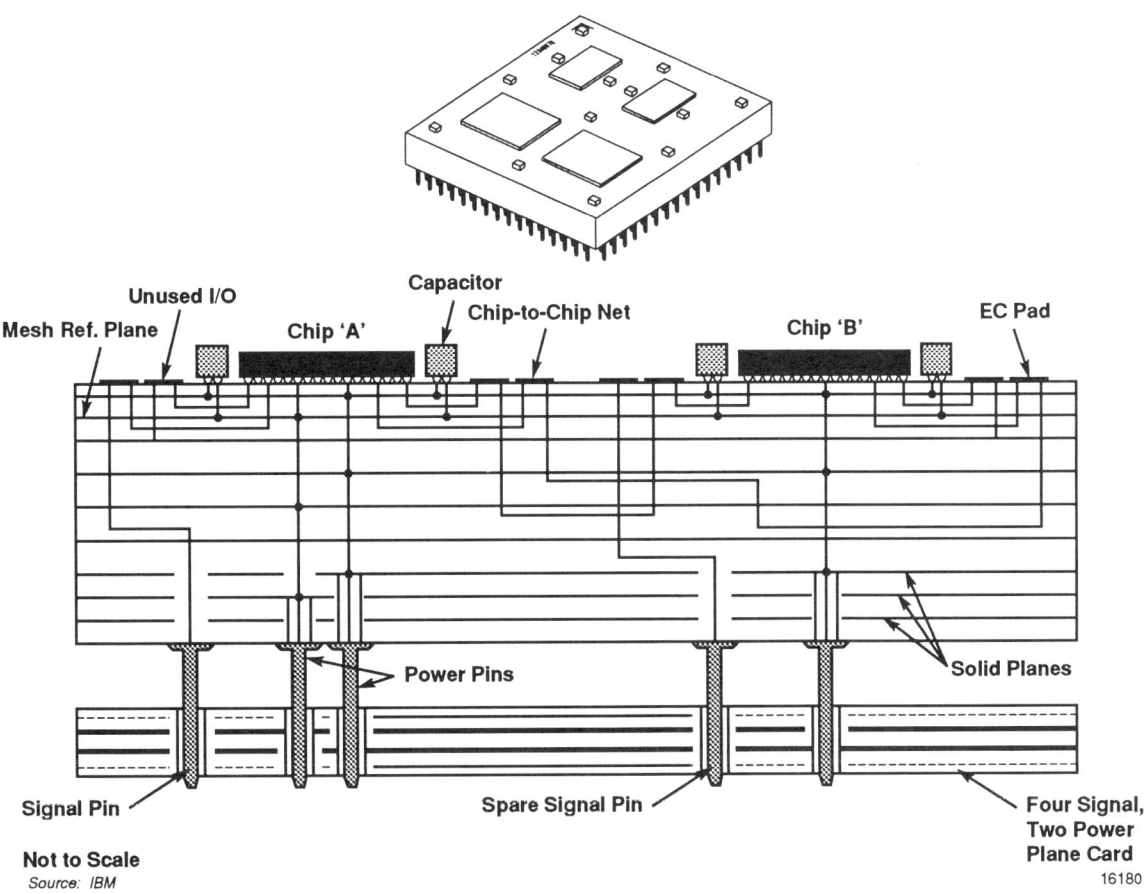

Figure 3-17. IBM MCM

INTEGRATED CIRCUIT ENGINEERING CORPORATION

Packaging

G. TAPE AUTOMATED BONDING (TAB)

Another approach that allows flexibility as well as high pin count is the use of area Tape Automated Bonding (TAB). TAB refers to the use of thin copper foil laminated to a plastic film, then etched (the copper) to form a pattern to match the bond pads on the IC chip and the pattern on the substrate. Figure 3-18 shows the application of area TAB to replace wire bonds in a PGA, but this technique also allows mounting the IC chip directly on an external ceramic or PCB substrate.

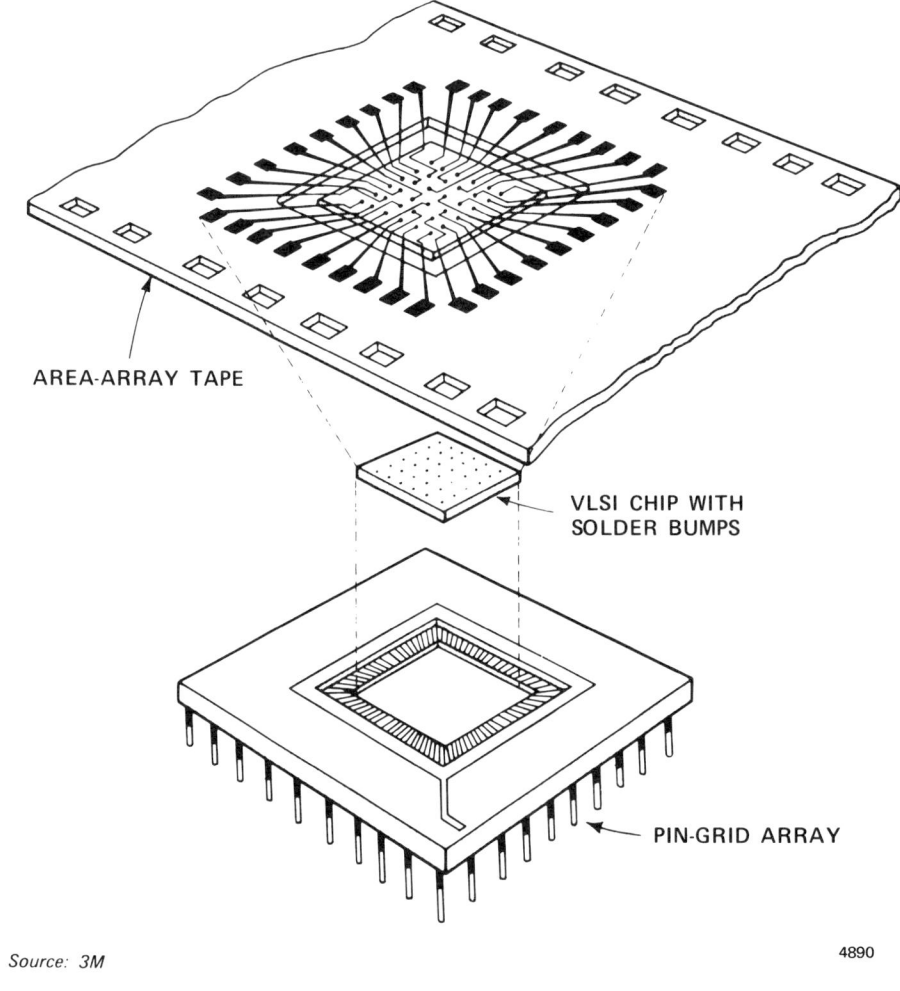

Source: 3M

Figure 3-18. Area Array TAB Concept

H. CHIP-ON-BOARD (COB)

IC chips can be mounted directly on the substrate, without a package. Figure 3-19 shows this approach with standard bonding techniques to form a hybrid circuit or MCM and Figure 3-20 shows a Chip On Board (COB) used in an electronic game cartridge.

Packaging

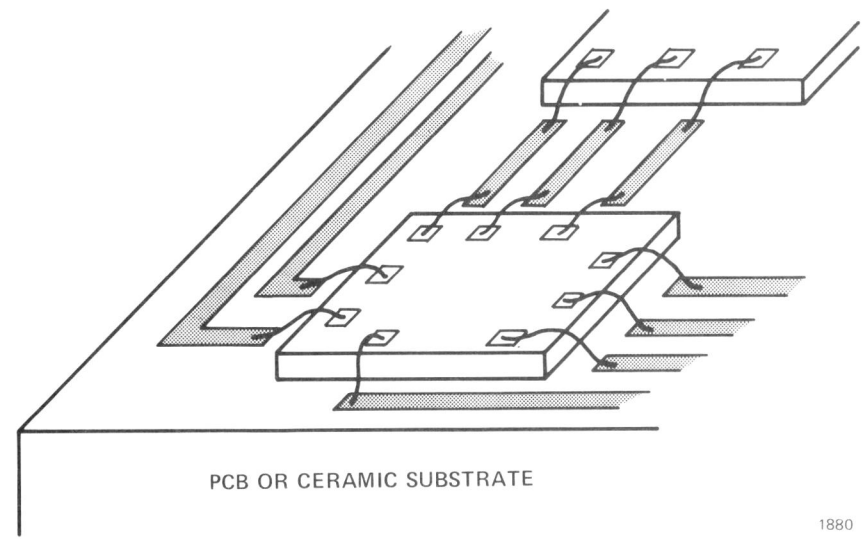

Figure 3-19. Hybrid Technologies Monolithic Chip-and-Wire

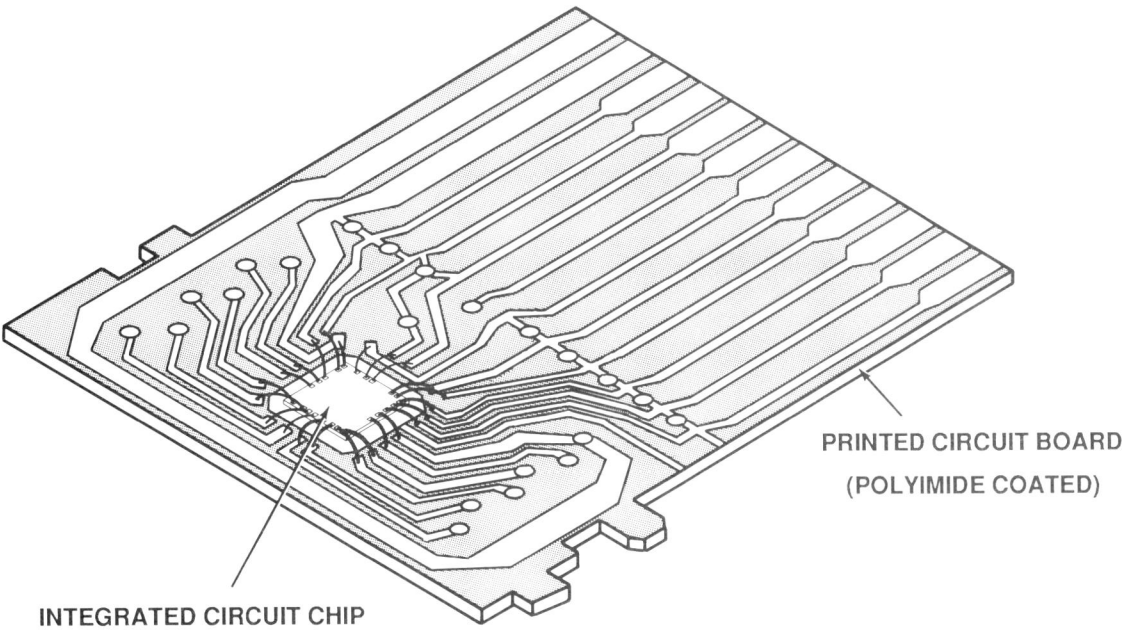

Figure 3-20. Chip-on-Board

I. SILICON SUBSTRATES

Another approach to reducing the thermal expansion problem is to construct MCMs on silicon substrates, then mounting the substrates onto ceramic or normal PCBs. This is shown in Figures 3-21 and 3-22.

Packaging

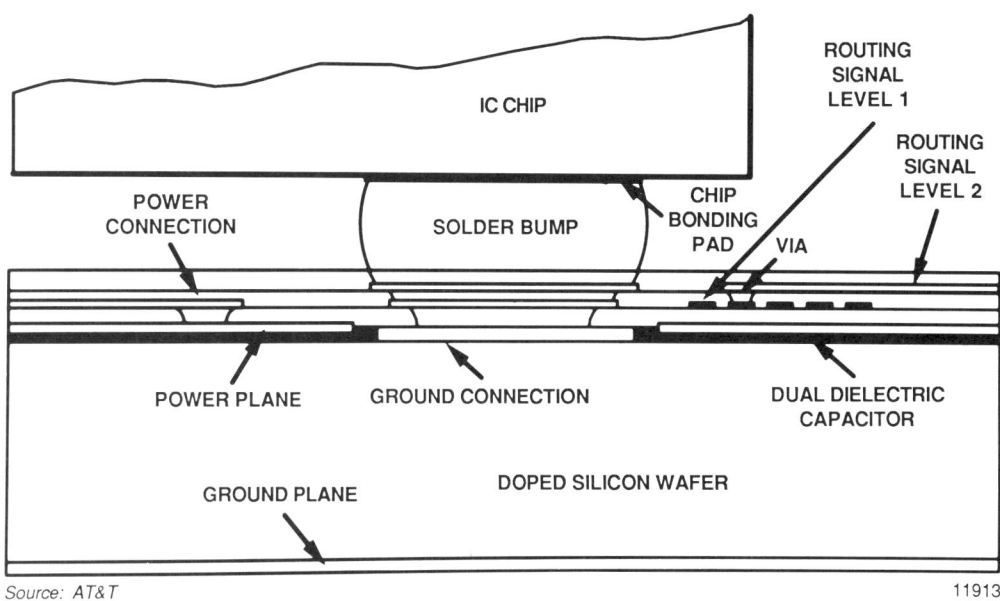

Figure 3-21. Interconnection Substrate and IC Chips

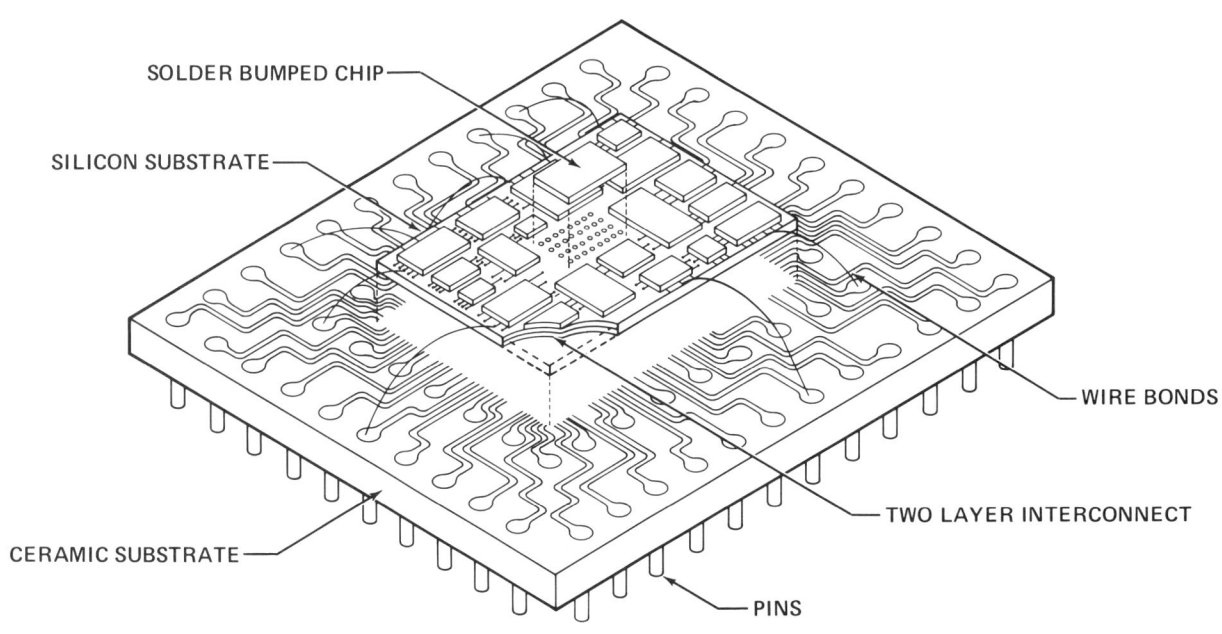

Figure 3-22. Silicon-on-Silicon Packaging Concept

J. BALL GRID ARRAY

Figure 3-23 shows a Ball Grid Array, which combines the high pin count available with the PGA with surface mount board assembly. In this technique the package is made of the same material as the substrate so the solder balls are sufficient to absorb the small thermal mismatch. In addition, the cost should be considerably lower than a standard PGA because the materials are less expensive. This package, or one very similar to it, will find wide use as a replacement for standard PGAs because of cost, size, and board density, as the ball grid can be made to finer pitch as the users become experienced with use of the package. If the pitch is reduced by a factor of two, the board area is reduced by a factor of four. A pitch of 20 mils is not unreasonable, compared to a pitch of 100 mils for the PGA. This is a theoretical size reduction of 25 times.

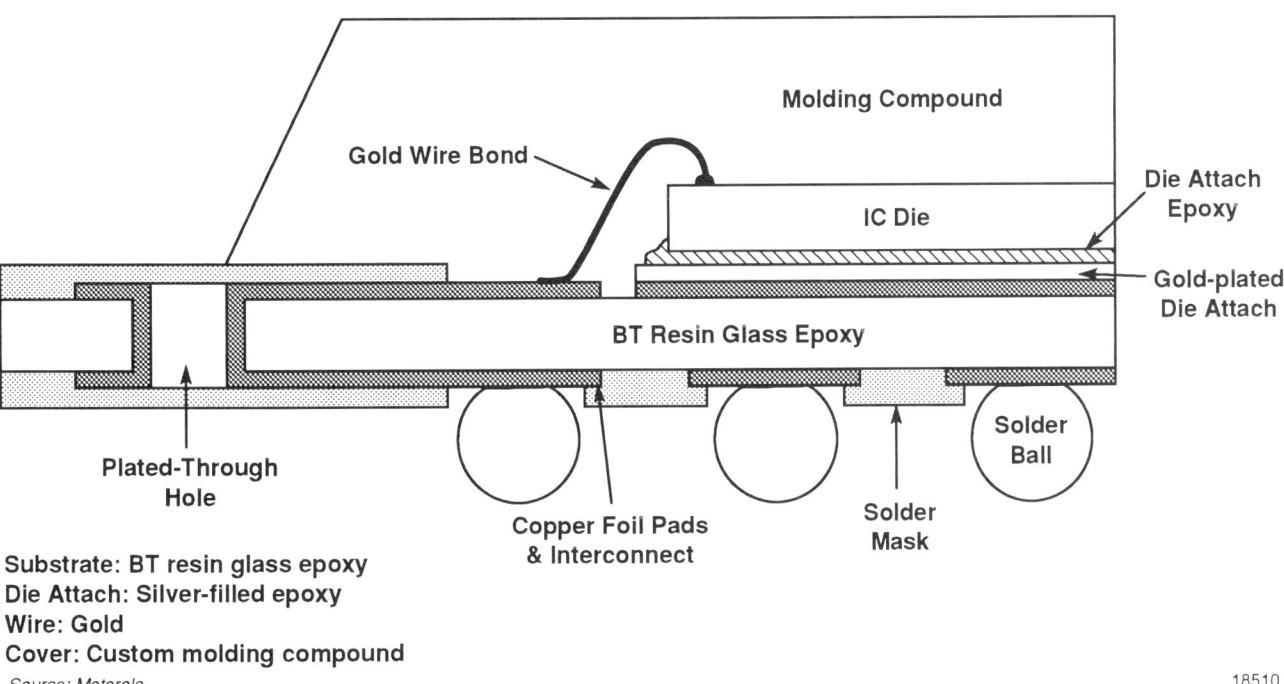

Figure 3-23. OMPAC Ball Grid Array From Motorola

K. SPECIAL PACKAGES

Special packaging is also important for high-volume ICs with special needs. An example of this is shown in Figure 3-24. The package is designed for memory chips where the die size is very large, the pin count is very low, and there are several chips used per system. The package is inexpensive, and its overall size is only slightly larger than the IC chip inside.

Packaging

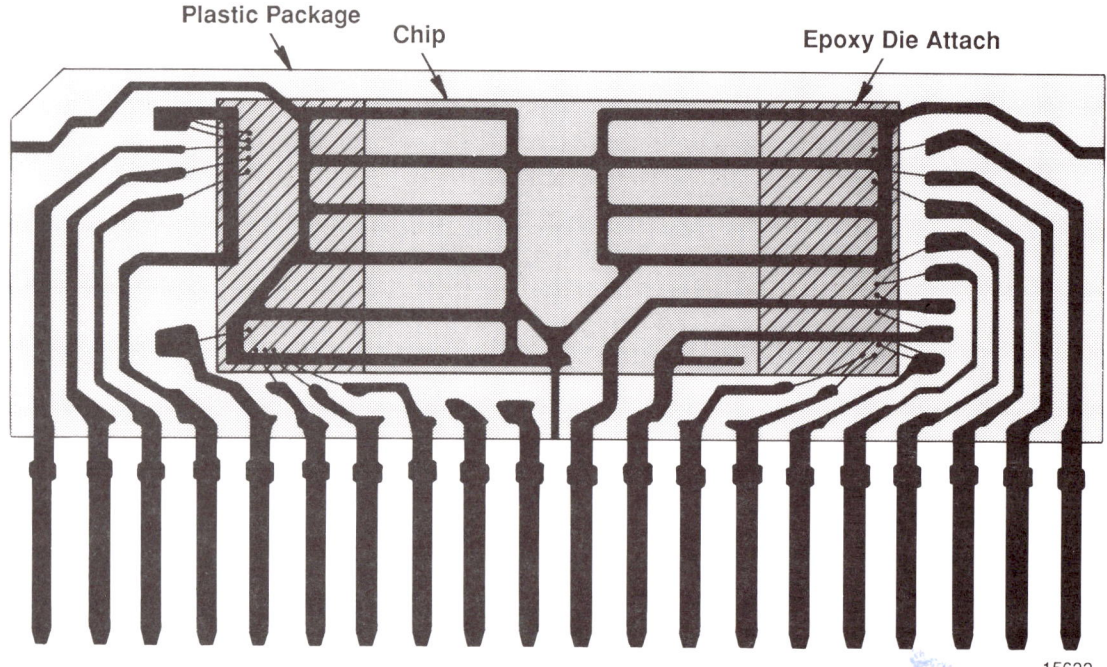

Figure 3-24. X-Ray of Hitachi 4M DRAM ZIP Package

The package in Figure 3-25 is constructed differently than most plastic devices. The IC chip itself is used as a base. The chip is covered with some type of coating and patterned to expose the bonding pads that run along the center. The leadframe is attached to the top of the IC chip and the bonding performed from the chip *up to* the leadframe. This package has the additional advantage of the ground and positive voltage supply terminals running almost the entire length of the chip.

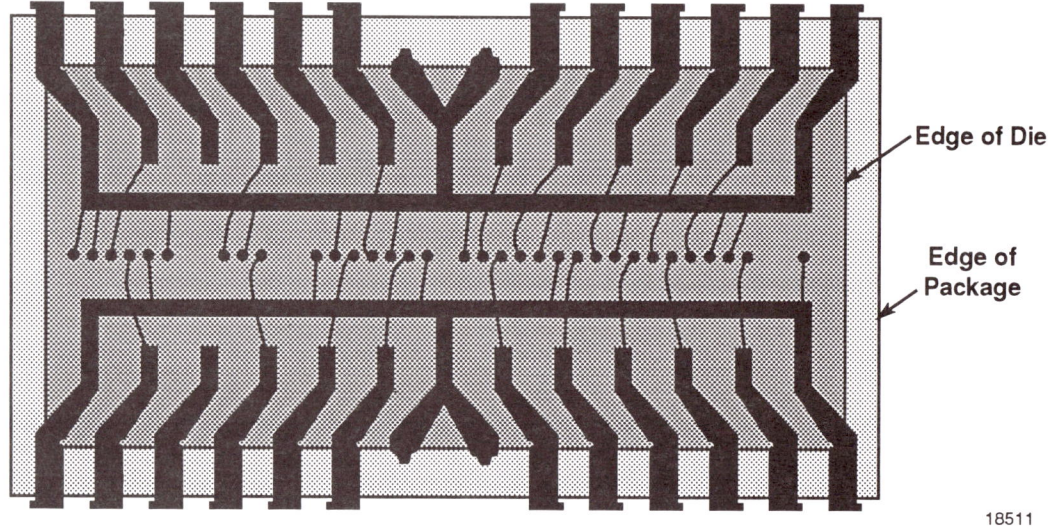

Figure 3-25. 16M DRAM

4 TERMS BY MAJOR AREA OF INTEREST

Semiconductor technology has its own vocabulary. The use of acronyms, jargon or "buzz words" often confuse our understanding of semiconductor technology. To keep the subject in the "basic" context, this section (chapter) will list many of the terms used throughout the book to familiarize the reader with terms of general science. Knowledge of these important terms will enhance the understanding of the subject.

The United States is the only major country in the world using the English system of units. All other countries use the metric system. Therefore, terms will be given in both dimensional systems.

A. TABLES

1. Metric Unit Prefixes (Figure 4-1).
2. Length Conversion Factors (Figure 4-2).
3. Length, Area, Mass Conversion Factors (Figure 4-3).
4. Silicon Wafer Summary (Figure 4-4).
5. Electrical Units (Figure 4-5).
6. Periodic Table of the Elements (Figure 4-6).

B. KEY SEMICONDUCTOR-RELATED ELEMENTS

- Aluminum
- Antimony
- Arsenic
- Boron
- Compound Semiconductor (III - V)
    GaAs
- Germanium
- Phosphorous
- Silicon
- Tungsten
- Titanium
- Gold
- Silver
- Platinum

C. KEY SEMICONDUCTOR-RELATED CHEMICALS

1. Acids

- Acetic (CH_3OOH)
- Buffered Oxide Etch (BOE) (NH_4/HF)
- Hydrofluoric (HF)
- Hydrogen Peroxide (H_2O_2) (Oxidizer)
- Nitric (HNO_3)
- Phosphoric (H_3PO_4)
- Sulfuric (H_2SO_4)

2. Bases

- Ammonium Fluoride (salt)
- Ammonium Hydroxide
- Potassium Hydroxide
- Sodium Hydroxide

3. Solvents

- Acetone (CH_3COCH_3)
- Alcohol, Reagent (CH_3OH)
- Ethylene Gycol (CH_2OHCH_2OH)
- Freon
- Hexamethyl disiliane (HMDS)
- Isopropyl Alcohol ($CH_3CHOHCH_3$)
- Methanol (CH_3OH)
- N-butyl Acetate ($CH_3COOCH_2CH_2CH_2CH_3$)
- Negative Resist
- Negative Resist Developer
- Positive Resist
- 2-Propanal ($CH_3CHOHCH_3$)
- Xylene ($C_6H_4CH_3)_2$
- Positive Photoresist
- 1, 1, 1, Trichloroethane (CCl_3CH_3)

4. Gases – General

- Argon
- Hydrogen
- Nitrogen
- Oxygen

5. Gases/Liquids – Process

- Arsine (A_3H_3)
- Diborane (B_2H_6)
- Dichlorosilane (SiH_2Cl_2)
- Hydrogen Chloride (HCl)
- Phosphine (PH_3)
- Phosphorus Oxychloride ($POCl_3$)
- Boron Tribromide (BBr_3)
- Boron Trifluoride (BF_3)
- Boron Trichloride (BCl_3)
- Silane (SiH_4)
- Trichlorosilane ($SiHCl_3$)
- Silicon Tetrachloride ($SiCl_4$)
- Ammonia (NH_3)
- Chlorine (Cl_2)
- Nitrous Oxide (N_2O)
- Nitrogen Oxide (NO_2)
- Tetraethylorthosilicate [$Si(OC_2H_5)_4$]

Terms by Major Area of Interest

Prefix	Symbol	Power of 10	Numerical Value	
tera	T	10^{12}	trillion	1,000,000,000,000
		10^{11}	hundred-billion	100,000,000,000
		10^{10}	ten-billion	10,000,000,000
giga	G	10^{9}	billion	1,000,000,000
		10^{8}	hundred-million	100,000,000
		10^{7}	ten-million	10,000,000
mega	M	10^{6}	million	1,000,000
		10^{5}	hundred-thousand	100,000
myria	my	10^{4}	ten-thousand	10,000
kilo	k	10^{3}	thousand	1,000
hecto	h	10^{2}	hundred	100
deka	da	10^{1}	ten	10
		10^{0}	one	1
deci	d	10^{-1}	tenth	.1
centi	c	10^{-2}	hundredth	.01
milli	m	10^{-3}	thousandth	.001
		10^{-4}	ten-thousandth	.000 1
		10^{-5}	hundred-thousandth	.000 01
micro	μ	10^{-6}	millionth	.000 001
		10^{-7}	ten-millionth	.000 000 1
		10^{-8}	hundred-millionth	.000 000 01
nano	n	10^{-9}	billionth	.000 000 001
		10^{-10}	ten-billionth	.000 000 000 1
		10^{-11}	hundred-billionth	.000 000 000 01
pico	p	10^{-12}	trillionth	.000 000 000 001
		10^{-13}	ten-trillionth	.000 000 000 000 1
		10^{-14}	hundred-trillionth	.000 000 000 000 01
femto	f	10^{-15}	quadrillionth	.000 000 000 000 001
		10^{-16}	ten-quadrillionth	.000 000 000 000 000 1
		10^{-17}	hundred-quadrillionth	.000 000 000 000 000 01
atto	a	10^{-18}	quintillionth	.000 000 000 000 000 001

Figure 4-1. Metric Unit Prefixes

Unit	Quantity in Equivalent Units					
	inch	mil	cm	mm	µm	Å
inch	1	10^3	2.54	25.4	2.54×10^4	2.54×10^8
mil	10^{-3}	1	2.54×10^{-3}	2.54×10^{-2}	25.4	2.54×10^5
cm	0.3937	3.937×10^2	1	10	10^4	10^8
mm	3.937×10^{-2}	39.37	0.1	1	10^3	10^7
µm	3.937×10^{-5}	3.937×10^{-2}	10^{-4}	10^{-3}	1	10^4
Å	3.937×10^{-9}	3.937×10^{-6}	10^{-8}	10^{-7}	10^{-4}	1

Figure 4-2. Length Conversion Factors

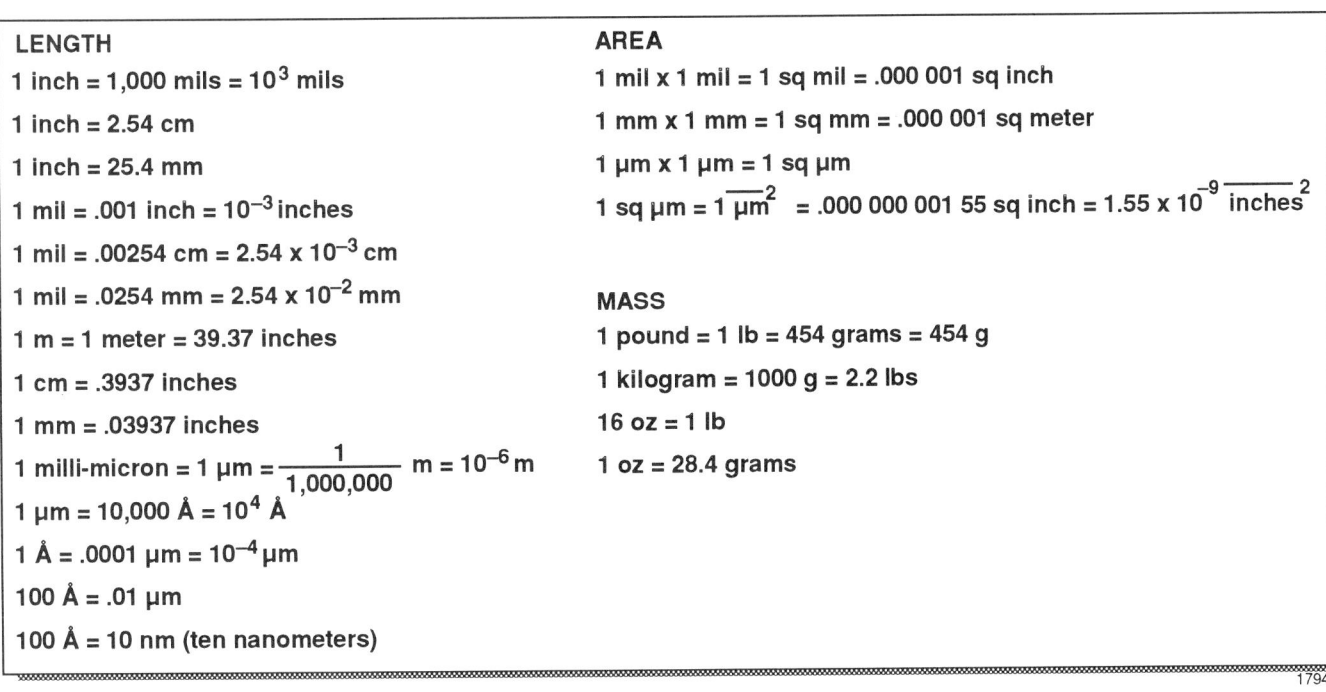

LENGTH

1 inch = 1,000 mils = 10^3 mils

1 inch = 2.54 cm

1 inch = 25.4 mm

1 mil = .001 inch = 10^{-3} inches

1 mil = .00254 cm = 2.54×10^{-3} cm

1 mil = .0254 mm = 2.54×10^{-2} mm

1 m = 1 meter = 39.37 inches

1 cm = .3937 inches

1 mm = .03937 inches

1 milli-micron = 1 µm = $\frac{1}{1,000,000}$ m = 10^{-6} m

1 µm = 10,000 Å = 10^4 Å

1 Å = .0001 µm = 10^{-4} µm

100 Å = .01 µm

100 Å = 10 nm (ten nanometers)

AREA

1 mil x 1 mil = 1 sq mil = .000 001 sq inch

1 mm x 1 mm = 1 sq mm = .000 001 sq meter

1 µm x 1 µm = 1 sq µm

1 sq µm = $1 \overline{µm}^2$ = .000 000 001 55 sq inch = $1.55 \times 10^{-9} \overline{\text{inches}}^2$

MASS

1 pound = 1 lb = 454 grams = 454 g

1 kilogram = 1000 g = 2.2 lbs

16 oz = 1 lb

1 oz = 28.4 grams

Figure 4-3. Length, Area, Mass Relationships

Wafer Size	Thickness		Area		Weight	
mm/in	mils	microns	in²	cm²	grams	ounces
50.8/2	11.0	279	3.14	2,025.80	1.32	0.047
76.2/3	15.0	381	7.07	4,561.28	4.05	0.143
100/3.94	20.7	525	12.19	7,864.50	9.67	0.341
100/3.94	24.6	625	12.19	7,864.50	11.45	0.404
125/4.92	24.6	625	19.02	12,271.85	17.87	0.630
150/5.9	26.6	675	27.39	17,671.46	27.82	0.980
200/7.87	28.5	725	48.69	31,415.93	52.98	1.870

Silicon density = 2.33 grams/cm³

Figure 4-4. Silicon Wafer Summary

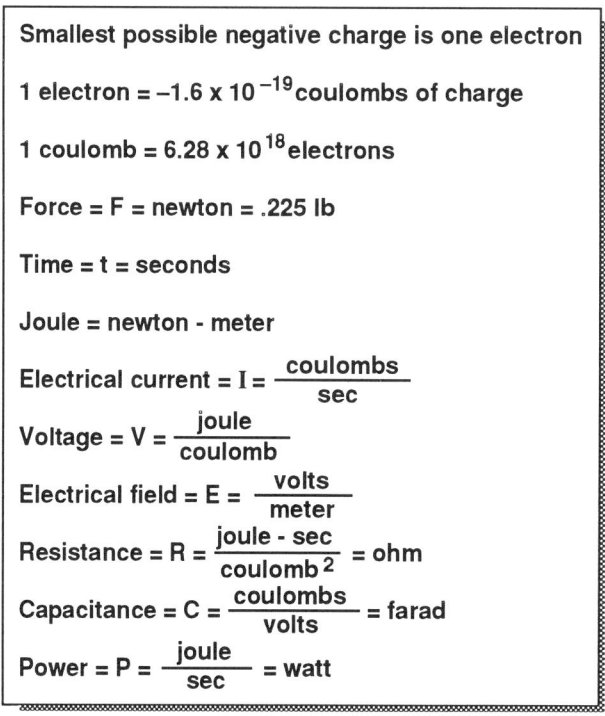

Figure 4-5. Electrical Units

Terms by Major Area of Interest

Figure 4-6. Periodic Table of the Elements

GLOSSARY

Acceptor	An impurity that can make a semiconductor P-type by causing the absence of electrons in the conduction band (called "*holes*"). These "holes" are carriers of positive charge. See donor.
Active Component	A (non-mechanical) circuit component that has gain or switches current flow, such as a diode, transistor, etc.
Aligner	An optical system used in transferring a mask or reticle image to a wafer.
Alignment	The arranging of a mask and wafer in correct positions with respect to each other. After alignment, radiation-sensitive photoresist on the wafer is exposed by radiation passing through the non-opaque areas of the mask.
Alignment Mark	A reference mark used in the alignment of the several photomask layers required for a single device or circuit.
ALSTTL	Advanced LSTTL. A fast bipolar logic family with power dissipation about half that of LSTTL.
Alloy	(1) In semiconductor processing the alloy process step refers to a heat treatment used to improve the metallurgical interaction between the silicon substrate and the interconnect metals. This improves ohmic contact. (2) Metallurgically is used to mean the blending together of two or more metals to form a particular compound.
Alternating Current	Electrical current that reverses (or alternates) at regular intervals. Abbreviated AC.
Aluminum (Al)	The metal often used in semiconductor technology to form the interconnects between devices on a chip. It is usually deposited by evaporation or sputtering. Aluminum may also be used as a P-type dopant.

Glossary

Amorphous	An atomic structure having no definite or recognizable form.
Amplifier	A device that uses an active component to increase the voltage or power of a signal without distorting its waveshape.
Analog	A continuous, non-digital representation of phenomena. An analog voltage, for example, may take any value.
AND Gate	A gate whose output is ON only if all input signals are ON.
Angstrom (Å)	A unit of length. 1/10,000 of a micrometer ($10^{-4}\mu m$).
Anisotropic	In a subtractive process, the material is removed only in the vertical direction in such a way that the side of the cut is perpendicular to the bottom of the cut.
Anneal	To heat a material to some elevated temperature to reorder the crystal structure and remove stresses. The subsequent cooling schedule affects the final level of stresses.
Antimony (Sb)	An N-type dopant often used to form the buried layer in a bipolar structure.
Array	A group of items (elements, leads, bonding pads, circuits, etc.) arranged in rows and columns.
Arsenic (As)	An N-type dopant often used to form the buried layer in a bipolar structure. Also used as an implant source in both bipolar and MOS processing.
Ashing	A method of stripping photoresist by a plasma (an electromagnetically-excited reactant gas).
Assembly	The final stage of semiconductor manufacturing, where the active device is encased in a plastic, ceramic or metal package. Also referred to as "*back-end*" processing.
Atmospheric Oxidation	The process of oxidizing silicon with some species of oxygen in a process tube at atmospheric pressure.
Atomic Number	A number assigned to each element that identifies the number of protons in the nucleus of an atom.

Autodoping	Atoms that outgas from a material and then are reabsorbed into the process ambient and materials.
Automation	The art of making processes or machines self-acting or self-moving. Also pertains to the technique of making a device, machine, process or procedure more fully automatic.
BTAB	(Bumped Tape Automated Bonding). A bonding process using inverse beams with solder (or other) bumps on the beams instead of on the chips. Obviates the need for special bumped chips.
Ball Bonding	See Bonding, Ball
Base	The control portion of a bipolar transistor. In an NPN transistor, the P-type material forms the base.
Base Diffusion	The diffusion during which the base regions of transistors are formed.
Batch	A number of wafers processed as a group.
Beam Leads	Thick, strong leads deposited directly on an integrated circuit chip and used for interconnecting the circuit into the system.
Beam Tape	Polyimide tape supporting copper foil shaped into beam leads for TAB bonding. Specifically designed for in-line automation of IC packaging.
Binary	A system of numbers using 2 as a base, in contrast to the decimal system which uses 10 as a base. The binary system requires only two symbols: 0 and 1.
Bipolar	An electronic device whose operation depends on the transport of both holes and electrons.
Bit	A binary digit. A bit is the smallest unit of storage in a digital computer, and is used to represent one of the two states in the binary number system.
Boat	A wafer holder made from quartz or polycrystalline silicon for use in furnace operations of semiconductor fabrication. Also may be made of teflon for transporting wafers between processing locations.

Glossary

Boat Puller	A electro-mechanical device interfaced to the oxidation/diffusion furnaces for inserting or removing at a given rate a number of boats loaded with wafers.
BOE	(*Buffered Oxide Etch*). A hydrofluoric acid (HF) and ammonium fluoride (NH_4F) solution used to etch silicon dioxide.
Bonding, Ball	A thermal compression bonding technique used only with gold wire. The wire end is melted to form a ball, which provides a larger area of contact than otherwise possible.
Bonding, Die	Attaching of the semiconductor die to the package substrate, with epoxy adhesives, gold eutectic or solder alloy. Also called die attachment.
Bonding Pads	Comparatively large metallization areas usually placed around the perimeter of the integrated circuit die to provide the areas to which wires from internal terminations of the leads of the package are connected.
Bonding Wedge	A form of thermal compression bonding used for microelectronic assembly, so named because the bond shapes the wire in a wedge shape.
Bonding Wire	Fine wires, usually aluminum or gold, connecting the metal bonding pads on an integrated circuit to the internal terminations of the leads of the package.
Boolean Algebra	A logical calculus named for mathematician George Boole, using alphabetic symbols to stand for logical variables, and 0 and 1 to represent states. AND, OR, and NOT are the three basic logic operations in this algebra. NAND and NOR are each combinations of two of the three operations.
Boron (B)	The P-type dopant commonly used for the isolation and base diffusion in standard bipolar NPN IC processing, and source/drain regions in PMOS transistors.
Boron Trichloride (BCl_3)	A gaseous mixture of boron and chlorine. The mixture is used as a dopant source for P-type diffusions and as an etchant gas for etching aluminum and aluminum alloys.

Glossary

Boule	A term used to describe the single-crystal silicon mass after the crystal growing process. Also called ingot.
Brazing	The process of joining two or more metals by partial fusion with a layer of hard soldering alloy at high temperatures.
Breakdown (Junction)	A high carrier conduction condition arrived at as a result of the field (voltage) being sufficient to cause this high level of conductance.
Bubbler	A container holding a liquid through which some type of inert carrier gas is passed to carry some partial pressure of the liquid into a process tube or a reaction chamber.
Bumped Chip	A chip from a wafer that has been specifically processed with buffer metal(s) over the I/O pads, followed by an addition of solder or gold "bumps" to provide reflow or thermocompression bonding areas for copper beam-lead attachment. See TAB.
Buried Layer	A low-resistivity, diffused region placed under the collector of a bipolar transistor to reduce its series resistance, commonly employed with an epitaxial structure.
Burn-In	Applying voltage to a semiconductor device at an elevated temperature for a given period of time to cause devices with marginal reliability to fail, eliminating potential field failures.
CAD	(*Computer-Aided Design*). A technique of using a computer to aid a person in the design of electrical circuits, integrated circuits, gate arrays, and other complex engineering designs in a reasonable timeframe.
Cantilever Loading	A mechanical support structure that allows process boats to be inserted and removed from the process chamber without touching the walls of the process chambers.
Capacitance (C)	The capability of storing electrical charge. Unit of measure is the Farad (F).
Capacitance - Voltage Plot or C-V Plot	(1) A measurement method used to characterize the dopant profile of a p-n junction. (2) An evaluation technique used to measure the quality of a dielectric for fixed charge density (Q_f), trapped charge density (Q_{ni}), and mobile charge density (Q_m). Depending on the methodology chosen, substrate characteristics can also be determined.

Carriers	Holes or electrons that are available in a semiconducting device for conduction of electric current.
Carrier Gas	An inert gas used to transport other elements to a process chamber or tube.
Cathode Sputtering	A method of depositing thin films, that employs a high-energy bombardment to release the source material.
CCD	Charge coupled device. A device utilizing a technique in which information is stored and transported by means of packets of minute electrical charges.
Centerless Grinding	A special grinding process used to shape the grown boule (ingot) into the final diameter. This process is done prior to slicing the boule into the individual wafers.
CERDIP	(*CER*amic *D*ual-*I*nline *P*ackage). A package assembled with the leadframe sandwiched between two ceramic layers and sealed by firing a glass frit.
CERMET	A combination of ceramic and metal powders used for thin-film resistors (and thick-film).
CERPACK	(*CER*amic *PACK*age). A cerdip-like package with the leadframe extending out all four sides, typically in surface-mounting format. Characteristics similar to cerdip. Also known as cerquad, cerpac or cerpak.
Channel	The conducting layer between source and drain of an MOS transistor that is induced by the applied gate voltage.
Chemical Etching	The use of liquid chemicals to remove a particular material.
Chip	An integrated circuit or discrete device. Also called a die.
Chip Carrier	A square (or rectangular) IC package with I/O connections on four sides; connections may be leadless or leaded.
Chrome, Chromium (Cr)	A metal element. (1) The metal is used as the opaque layer in a reticle and or mask. (2) The metal is used as part of a die attach alloy or metal interconnect composition.

Glossary

Circuit Layout	The physical arrangement of all the circuit elements on the surface of the device.
Cleanroom	An area specially constructed to control the air flow, temperature, and humidity in such a way that constant filtration keeps contamination below some predetermined level and termperature and humidity within predetermined limits.
Cleanroom Class	A government specification issued as FED-STD-209 that defined the number of particles of a given size and distribution per cubic volume of space for each class.
Coefficient of Diffusion	The rate at which a diffusant will diffuse into bulk material at a given temperature; measured in cm^2/sec.
CMOS	Complementary metal-oxide semiconductor. A logic family made by combining N-channel and P-channel MOS transistors.
Collector	The region of a bipolar transistor that "collects" the emitted electrons and then passes them on through a conductor, completing the electrical circuit.
Compound Semiconductor	A semiconductor usually formed by the compound of Group III and Group IV elements such as GaAs, GaInP, etc.
Concentration (Dopant)	The level of dopant materials as compared to "pure" semiconductor materials within the structure. The net concentration establishes the characteristic conduction pattern and other characteristics of the material.
Conductor, Electrical	A material capable of carrying (conducting) electricity. Silver is the best electrical conductor. Copper, gold, and aluminum are also popular conductors. Aluminum is the conductor most commonly used in IC fabrication.
Contact Aligner	An optical system that uses contact printing (the mask touches the wafer) to expose a wafer.
Contact Printing	Exposure of a wafer by passing light through a mask that is in direct contact with the photoresist-coated wafer. Chrome working plates are most commonly used due to their longer life, but emulsion working plates can be used to reduce damage to the wafer.

Contacts	The regions of exposed silicon that are uncovered prior to the metallization process to provide electrical access to individual devices.
Contamination	A general term used to describe unwanted material or foreign matter that adversely affects the electrical characteristics of a semiconductor wafer.
Critical Dimensions (CD's)	The width or space of critical circuit elements in an IC.
Crossovers	Locations in integrated circuits where separate current paths cross one another. These are accomplished by using a low resistance diffused path (emitter diffusion) for one path and the metallization for the other. The two paths are insulated from one another by the SiO_2.
Crystalography	The science and classification of crystals, particularly the semiconductor material employed to fabricate transistors and integrated circuits.
Crystal Orientation	The relationship of wafer surface to the crystal facets at which the crystal is sliced. Each crystal orientation has direct effect on device characteristics.
Cum (Cumulative) Yield	The output of several manufacturing process steps divided by the input or the yields of several consecutive steps multiplied together.
Current (I)	The flow of electrons. Usually measured in amperes (amp or A).
Curve Tracer	An electrical tester that displays a X & Y relationship on the face of a CRT.
CVD	(*Chemical Vapor Deposition*) A method of uniformly depositing films of insulators or conductors on a wafer to either isolate or connect circuit elements. Lithography defines a pattern of film that is to remain after an etch step.
Darlington Amplifier	An amplifier in which the collectors are tied together, and the emitter of the first directs current to the base of the second.
Deep Ultraviolet or Deep UV	Refers to the electromagnetic spectrum where the wavelength of light is less than 300 nanometers.

Glossary

Dehydration Bake	A heat cycle wafers receive prior to photolithography processing.
Depletion Device	A type of MOSFET which is "on" when no input signal is present.
Depletion Region	That area at a P-N junction which, when reverse biased, is swept clear of free charges.
Deposition	A heat or physical (sputtering) process whereby a thin film of material is deposited over the surface of a wafer.
Development	A chemical process that removes the photoresist from areas not defined by the mask in a lithographic procedure.
Develop Inspect or A.D.I.	The inspection a wafer receives after the develop process.
Developer	The liquid solution used to resolve an image after exposure.
Diborane (B_2H_6)	A gaseous compound of boron and hydrogen used as P-type dopant.
Die	A single square or rectangular piece of semiconductor material into which a specific electrical circuit has been fabricated. Plural is dice. Also called a chip or device (IC or discrete).
Dielectric	A non-conductor of current, an insulator
Die Sort	See Wafer Sort.
Die Bonding	See Bonding, Die.
Differential Amplifier	An amplifier that amplifies the voltage difference between two input signals and has two inputs and two outputs.
Diffusion	A high-temperature process by which selected chemicals (dopants) enter the crystalline structure of semiconductor materials to change the electrical characteristics at desired locations. This process takes place in a diffusion furnace.
Digital	A method of representing information in an electrical circuit by switching the current ON or OFF. Only two output voltages are possible, usually represented by "0" and "1."
Digital Circuit	A circuit that operates like a switch and can perform logical functions. Used in computers or similar logic-based equipment.

Glossary

Diode	A two-terminal device that allows current to flow in one direction but not the other. A diode is present at the intersection (junction) of a P-type and an N-type semiconductor.
D.I. (Deionized) Water	Water that has been specially processed to remove all ionized impurities and filtered to remove particle contamination.
DIP	(*Dual In-line Package*). The most common type of IC package; circuit leads or pins extend symmetrically outward and downward from the long sides of the rectangular package body.
Direct Current	The flow of electrons only goes in one direction. Abbreviated DC.
Direct Step On Wafer (DSW)	See Stepping Aligner
Discrete Device	A semiconductor containing only one active element, such as a transistor or a diode.
DMOS	(*Double-diffused MOS*). Also called VMOS (differing from V-groove MOS), for its vertical flow of current through the substrate.
Donor	An dopant that can make a semiconductor N-type by donating extra "free" electrons to the conduction band. The free electrons are carriers of negative charge. See acceptor.
Dopant (also Diffusant)	Dopants are the materials used to change the electrical characteristics of a semiconductor crystal, making it N- or P-type.
Doping	The introduction of an dopant into a semiconductor to modify its electrical properties by creating a concentration of N or P carriers. Doping is normally accomplished through diffusion or ion implantation processes.
Drain	The working current terminal (at one end of the channel in a MOSFET) that is the drain for holes or free electrons exiting from the channel. Corresponds to collector of a bipolar transistor.
Driver	An element that is coupled to the output stage of a circuit in order to increase its power, current handling capability, or fanout; for example, a clock driver is used to supply the current necessary for a clock line.

Glossary

Drive-In	The second part of a two-part diffusion. It is the part of the operation in which the diffusant deposited during predeposition is diffused further into the wafer to achieve the desired impurity profile.
Dry-Etch	The process that uses RF energy and gas phase chemicals to remove a specific layer during semiconductor processing.
Dry Oxide	The thermal oxidation of silicon using dry oxygen gas.
Dynamic Ram (DRAM)	A type of semiconductor memory in which the presence or absence of a capacitive charge represents the state (0 or 1) of a binary storage element. The charge must be periodically refreshed.
E-Beam	Electron beam. Refers to an e-beam evaporator or exposure system.
ECL	(*Emitter Coupled Logic*). A form of current-mode logic in which the output is available from an emitter-follower output stage.
EEPROM or E^2PROM	(*Electrically-erasable PROM.*) Similar to ROM, but with the capability of selective erasure and programming through special electrical stimulus. Sometimes termed EEPROM.
Electron	An elementary, negatively charged atomic particle.
E-Beam Evaporation	A deposition process that concentrates an electron beam upon a material to be evaporated, raising the material to its vaporization point. The material is deposited on the surface of the wafers as it cools (freezes).
E-beam Exposure System	An electronic imaging system that controls the exposure of a photo-imageable material by turning an electron-beam on and off, thus writing the exposure pattern of a database tape into the photoresist.
Edge Die	The two or three rows of dice along the outer circumference of the wafer.
Electromigration	The result of the movement of electrons in a conductor when the electric field is raised to a very high level and the operating temperature is high. This results in a physical displacement of conductor atoms. If the conditions are severe at high current densities, this eventually leads to an open conductor and circuit failure.

Ellipsometer	An optical measurement instrument that processes the optical constants of light and outputs the refractive index and thickness of films.
Emitter	The region of a bipolar transistor that serves as a source or input end for carriers. N-type for NPN, P-type for PNP.
Emitter Diffusion	The diffusion during which the emitters of the transistors are formed.
Emulsion Mask	A transparent mask plate on which the opaque region is a suspension of a salt of silver in gelatin or collodian.
Enchancement Device	A type of MOSFET that requires a control signal input to turn on the device. The device is "off" when no input signal is present.
Epitaxial Layer	A grown crystal layer (usually doped) having the same crystallographic orientation as the substrate (wafer). Also called "epi."
Epoxy	A family of thermosetting resins used in the packaging of semiconductor devices. Epoxies form a mechanical bond to many metal surfaces. Also used to attach dice to substrates or leadframes.
Epi	Short for epitaxy, the controlled growth of a layer of crystalline semiconductor material on a suitable substrate.
EPROM	(Erasable Programmable Read-Only Memory). Similar to ROM, but enables the user to erase stored information. Normally refers to a memory device whose contents may be erased by exposure to ultraviolet light shined through a window in the ceramic package.
Etch	The process of removing material (such as oxides or other thin films) by chemical, electrolytic or plasma (ion bombardment) means.
Evaporation	A vacuum process, usually at less than 10^{-6} Torr (mm of Hg), where metal(s) are vaporized through thermal agitation, then recrystallized on cooler surfaces, generally the semiconductor wafers.
Exposure	Subjecting a sensitive chemical or material to light or other radiant energy.
Fabrication	In semiconductor manufacturing, fabrication usually refers to the (front-end) process of making devices in semiconductor wafers, but usually does not include the package assembly (back-end) stages.

Glossary

Failure Analysis	An orderly procedure for determining the reason that a device has failed. The results are frequently useful for enhancing the reliability of subsequent products.
Failure Rate	A rate of failure per unit time, for example, 3 failures per 164 hours or 1 failure per day, etc. Semiconductors are usually measured in failures per 1000 hours.
Feature Size	See "Minimum Geometry".
Femto	A prefix meaning one quadrillionth ($\times 10^{-15}$). Symbol is f.
FET	(*Field Effect Transistor*). A solid-state device in which current is controlled between source and drain terminals by voltage applied to a gate terminal, which is insulated from the semiconductor substrate.
Field Oxide	A thick layer of dielectric, generally silicon dioxide, that covers the inactive portion of the semiconductor surface.
Final Test	The electrical evaluation of the packaged device.
Flat Band Voltage	The voltage at which the net charge in the semiconductor is zero.
Flat Pack	A ceramic surface-mounted hermetic package. Very popular for military applications because of its small size, but characterized by lead and package seal fragility, poor structural strength, excessive gold, and loose particle problems.
Flat Zone	The region in a process tub or chamber where the temperature is controlled to some specification tolerance.
Flip-Chip	An IC designed for face-down mounting, attached by controlled-collapse solder pillars on I/O (bonding) pads of the device.
Flip-Flop	An electrical circuit having two stable states: on and off. A basic logic circuit component.
Four-Point Probe	An electrical evaluation instrument containing four precisely spaced probes that contact the semiconductor surface to measure the dopants within the semiconductor wafer. This measurement is called sheet resistance.
FPLA	(*Field Programmable Logic Array*). A PLA that can be programmed by the user.

Frequency	The number of times per second an alternating current goes through a complete cycle. Formerly expressed in cycles per second. Now expressed in Hertz (Hz).
Furnace	A piece of process equipment containing heating elements connected to precision controller. The controller allows a zone of constant temperature to be maintained to a tight tolerance.
Fuse	An electrical circuit element that allows only some maximum level of current flow before it becomes an open circuit.
Fuseable Link	A circuit element used in memory circuits as a programmable connection.
GaAs	(*Gallium Arsenide.*) A compound semiconductor material in which active devices are fabricated. GaAs has a higher carrier mobility than silicon, thus it has the capability of producing higher speed devices.
GaAs FET	A discrete device used to amplify higher frequency radio signals.
Gate	(a) The basic digital logic element — where the binary value of the output depends on the values of the inputs. (b) The primary control terminal of a field effect transistor.
Gate Array	An IC consisting of a regular arrangement of gates that are interconnected to provide custom functions. Sometimes called an uncommitted logic array (ULA).
Gate Equivalent	The basic unit of measure for digital circuit complexity, based on the number of individual logic gates that would have to be interconnected to perform the same circuit function.
Gate Oxide	The thin layer of thermal oxide separating the gate electrode (terminal) from the semiconductor substrate.
Germanium (Ge)	A brittle, grayish-white metallic element having semiconductor properties. Widely used in crystal diodes and early transistors.
Giga	Prefix meaning one billion ($\times 10^9$). Symbol is G.
Glassivation	Passivation using silicon dioxide (glass).

Hard Surface	A photoplate coated with a relatively hard, non-emulsion coating that can be selectively etched to produce a photomask. Popular hard surface materials are chromium, chromium oxide, and iron oxide.
Hardware	The physical components of a computer or an electronic system — as opposed to software, which is the term used to describe the programs and instructions for a computer.
Heat Sink	An assembly that serves to dissipate, carry away, or radiate into the surrounding atmosphere heat that is generated by an active electronic device.
HEPA Filter	(*H*igh *E*fficiency *P*articulate *A*ir Filter). A specially constructed filter membrane that allows a high volume of air flow and stops small particles of some pre-determined size from passing through. In a semiconductor cleanroom the maximum particle size will range of 0.5μm down to <0.1μm.
High-Pressure Oxidation	A special pressure vessel contained inside a resistance heated furnace. This pressure vessel allows thermal oxidation to be done at raised pressure levels of 5 to 25 atmospheres.
Hillock	The displacement of a thin-film material that occurs when heated and cooled. Hillocks of aluminum are sometimes found in the aluminum interconnect metal on semiconductor devices.
Hexamethyldisilizane (HMDS)	A chemical primer used to promote improved adhesion between the photoresist layer and the underlying surface.
HMOS	(*High-performance MOS*). A scaled, high-performance NMOS structure.
Hole	The absence of a valence electron in a semiconductor crystal. The movement of a hole is equivalent to the movement of a positive charge.
Hybrid Circuit	A microelectronic device consisting of both film circuits and semiconductor elements.
Hydrofluoric Acid (HF)	A strong acid used to remove silicon dioxide. Generally it is mixed with ammouim fluoride for better etch-rate control.

Hydrogen (H_2)	A common gas used in semiconductor processing. One must be cautious of the explosive nature of hydrogen.
Image	The defined pattern on a surface.
Integrated Circuit (IC)	A semiconductor die containing multiple elements that act together to form the complete device circuit.
I^2L	(*Integrated Injection Logic.*) A bipolar structure characterized by an integrated PNP load device and inverted operation of an NPN logic transistor.
Impurity	Any foreign or undesired material that gets incorporated into the semiconductor material or layers above the silicon surface.
Ingot	A term used to describe the single-crystal silicon mass after the crystal growing process. Also called boule.
Interconnect	A conductive connection between two or more circuit elements. The conductors among elements (transistors, resistors, etc.) on an integrated circuit or between components on a printed circuit board.
Insulator	A material that is a poor conductor of electricity — used to separate conductors from one another or to protect personnel from electricity.
Ion	An atom that has either gained or lost electrons making it a charged particle (either negative or positive, respectively).
Ion Beam Milling	A process of physically removing unwanted (unprotected) material from a semiconductor surface by ion bombardment.
Ion Implantation	A means for adding dopants to a semiconductor material. Charged atoms (ions) are accelerated in an electric field into the semiconductor material. It is especially useful for thin doped areas. This process is much more precise than the diffusion method of doping.
Isolation Diffusion	A method of achieving isolation by diffusing in such a manner that P-N junctions surround the areas to be isolated from one another.
Isotropic	In a subtractive process, the material is removed in all directions simultaneously, frequently at the same rate vertically and horizontally.

Glossary

ISL	*Integrated Schottky Logic*.
JFET	(*Junction Field Effect Transistor*). A solid-state device in which current is controlled between source and drain regions by voltage applied to a conducting or junction gate terminal.
Junction (P-N)	The line of demarcation where the number of P- and N-type carriers are exactly equal with a surplus of P-type on one side and N-type on the other.
Kilo	A prefix meaning a multiple of 1000 ($\times 10^3$). Symbol is K.
Latch Up	(1) In an electrical circuit this refers to when the circuit is conducting and hold this state. (2) In CMOS this refers to the parasitic conduction that occurs when the parasitic n-p-n and p-n-p transistors get turned on.
Laterial PNP Transistor	An integrated transistor that is made in such a way that the transistor action is laterally across the surface of the diffusions rather than vertically through the N and P diffusions.
Leadframe	A stamped or etched metal frame that provides external electrical connections for a packaged electrical device.
LCC	(*Leadless Chip Carrier*). Usually made of ceramic material.
LED	(*Light-Emitting Diode*). A compound semiconductor device that emits light whenever current passes through it.
Line Width	Width of an opaque line. Usually refers to a dimension on a mask or a feature on an IC.
Linear Device	An amplifying-type (or analog) device, as opposed to digital device.
Linear Circuit	A circuit whose output is an amplified version of its input or whose output is a predetermined variation of its input.
Lithography	The transfer of a pattern or image from one medium to another, as from a mask to a wafer. If light is used to effect the transfer, the term, "*photolithography*" applies. "*Microlithography*" refers to the process as applied to images with features in the submicron range.

INTEGRATED CIRCUIT ENGINEERING CORPORATION

LSI	(*Large-Scale Integration*). ICs containing 1,000 or more transistors, but less than 100,000.
LSTTL	(*Low-power Schottky TTL*). The power dissipation of LSTTL is typically one-fifth that of conventional TTL.
Magnetron Reactive Ion Etch (MRIE)	The use of magnet to shape the ion field of etchant molecules in a reactive ion etch system.
Majority Carrier	The mobile charge that predominates in a semiconductor material. Electrons in N-type material and holes in P-type material.
Mask	A transparent (glass or quartz) plate covered with an array of patterns used in making integrated circuits. Each pattern consists of opaque areas that prevent light passage. The mask is used to expose photoresist that defines areas to be later etched on a wafer. Masks may use emulsion, chrome, iron oxide, silicon, or other material to produce the opaque areas.
Mask Trimmer	A device that uses bursts of laser energy to vaporize and remove unwanted opaque areas or spots on a photomask or reticle.
Mass Flow Controller (MFC)	A gauge consisting of a mass flow meter, an electronic controller and a valve that can control the rate of flow of a gas.
Master Plate	Original photomask generated by a step-and-repeat camera or e-beam system, used to print submasters or working plates.
MCBF	(*Mean Cycles Between Failures*) is obtained by dividing Mean Time Between Failures by the length of time required to complete one operating cycle.
Mega	A prefix meaning a multiple of one million ($\times 10^6$). Symbol M.
Memory	A general term for computer hardware that holds information in electrical or magnetic form.
MESFET	(*Metal Schottky FET*). A field effect transistor whose gate structure consists of a metallic Schottky barrier.
Metal-Gate MOS	MOS devices with a metal layer deposited to form the gate elements.

Metal Mask	The photomask used to define the metal interconnects in an IC.
Metallization	Metallization is the process of depositing a thin film of metal and patterning it to form the desired interconnection arrangement.
Micro	A prefix meaning one-millionth (x10^{-6}). Symbol is μ
Microcomputer	A microprocessor complete with stored program memory (ROM), random-access memory (RAM), and input/output (I/O) logic. Microcomputers are capable of performing useful work without additional supporting logic. If all functions are on the same chip, this is sometimes called a *microcontroller*.
Micrometer	One-millionth (10^{-6}) of a meter. About 40 millionths of an inch. Synonymous with micron. Symbol is μm.
Micron	A term used for micrometer.
Microprocessor	The basic arithmetic logic of a computer. Also called MPU for microprocessor unit.
Mil	One-thousandth of an inch (10^{-3} inch). Equal to 25.4 microns.
Miller Indices	A set of three integers used to define a set of parallel planes. The integers are derived from the coordinate system of the basic vectors.
Milli	A prefix meaning one-thousandth (x10^{-3}). Symbol is m.
Minority Carrier	The non-predominant mobile charge carrier in a semiconductor. Electrons in P-type material and holes in N-type material.
Minimum Geometry	The smallest line width or spacing between lines or features on a semiconductor die. May be defined "as drawn" or "as printed."
MMIC	(*M*onolithic *M*icrowave *IC*). An analog IC with on-chip inductors and capacitors designed to work at microwave frequencies.
MNOS	(*M*etal *N*itride *O*xide *S*emiconductor). The dielectric between metal and semiconductor is fabricated from silicon nitride. This technology is commonly used to make EEPROMs.
Mobility (Hole or Electron)	The rate with which a carrier (majority or minority) moves within a material under the influence of an electric field.

Glossary

Molecule	An aggregate of two or more atoms of a substance that exists as a unit.
Monochromatic Light	A single wavelength of light energy.
Monolithic Device	A device whose circuitry is completely contained on a single die or chip.
MOS	(*Metal Oxide Semiconductor*). Refers to field effect transistors (FETs) in which current flows through a channel of semiconductor material is controlled by the electric field under a gate structure.
MOSFET	A type of field effect transistor. See MOS.
MPU	(*Microprocessor Unit.*) Sometimes used synonymously with the word microprocessor.
MSI	(*Medium-Scale Integration*). A term generally applied to integrated circuit chips containing 30 or more transistors, but less than 1,000.
MTBF	(*Mean Time Between Failures*) is the reciprocal of the sum of the failure rates of every component in a system.
NAND Gate	(*Not-AND* gate). An AND gate followed by an inverter. The output of the AND gate is inverted to the opposite value.
Nano	A prefix meaning one-billionth ($\times 10^{-9}$). Symbol is n.
Negative Resist	A photoresist that remains in areas that are exposed by light shined through the clear regions of a mask. A negative image of the mask remains following the development procedure.
Nitric Acid (HNO_3)	A strong acid used in semiconductor processing.
Nitrogen (N_2)	An inert gas used as a carrier or purge gas in semiconductor processing.
NMOS	Also called *N-channel MOS*. A type of MOSFET using electrons to conduct current in the semiconductor channel. The channel has a predominantly negative charge during conduction. The source and drain are N-type.
Noise Immunity	A measure of the insensitivity of a logic circuit to triggering by or reaction to spurious or undesirable electrical signals or noise, largely determined by the signal swing of the logic.

NOR Gate	(*Not-OR.*) An OR gate followed by an inverter. The output of the OR gate is inverted to the opposite value.
NOT Gate	The output is just the opposite from the single input.
N-type Semiconductor	A semiconductor crystal containing a small amount of "*dopant*" atoms that have one more valence electron that the other atoms in the crystal. These extra negative electrons can find no unoccupied bonds to bind them, so they are free to wander and constitute electric current. Common N-type dopants for silicon are phosphorus and arsenic.
OP Amp	(*Operational amplifier*). A general-purpose IC used as a basic building block for implementation of linear functions.
Opaquing	The applying of minute droplets of opaquing ink to fill pinholes and correct geometric areas of mask or reticles.
OR Gate	The output is yes if at least one input is yes.
Oxidation	The chemical process of joining oxygen with another element. In semiconductors, oxidation means the joining of oxygen and silicon to form silicon dioxide (SiO_2).
Oxide	Usually refers to silicon dioxide (SiO_2) in semiconductor terminology.
Oxide Etching	The removal of silicon dioxide.
Oxygen (O)	An element that is used in the formation of silicon dioxide (SiO_2).
Package	The protective container for an electronic component — with terminals to provide electrical access to the components inside.
Parasitic	A parasitic is an undesirable stray capacitance, inductive coupling, or resistance leakage, as well as undesired transistor actions. The first and last are most serious in monolithic integrated circuits.
Parfocal	The field of view essentially remains in focus when switching between objective lens on a microscope.
Parts Per Million	The presence of one part in a million parts.

Passivation	A layer of material put over a wafer to stabilize and protect the surface. Silicon dioxide or silicon nitride are often used for IC passivation.
Passive Components	An electrical component without "gain" or current-switching capability. Commonly used when referring to resistors, capacitors or inductors.
Patterning	The transferring of an image from a reticle or mask to the photo resist.
Pattern Generator	A system that automatically generates a reticle from the coded images defined by a CAD system.
Pellicle	A thin, transparent membrane that seals off the mask or reticle surface from airborne contamination.
PG Tape	The magnetic tape output of a CAD system that is used to drive a pattern generator.
Phosphine (PH_3)	A compound of phorphorus and hydrogen used in semiconductor processing. The material is very toxic. It is a N-type dopant.
Phosphorus (P)	The N-type dopant commonly used for the buried layer contact and emitter diffusions in standard bipolar IC technology, and the source/drain regions in NMOS.
Phosphorus Oxychloride ($POCl_3$)	A chlorinated phosphorus compound used as an n-type dopant.
Photolithography	Lithographic techniques involved light as the pattern transfer medium. See Lithography.
Photoplate	The term used for a mask before images have been formed on it.
Photomask	See Mask.
Photoresist	A light-sensitive liquid that is spread as a uniform thin film on a wafer or substrate. After baking, exposure of specific patterns is performed using a mask. Material remaining after development resists subsequent etch or implant operations.
Pico	A prefix meaning a trillionth or one-millionth of one-millionth ($\times 10^{-12}$). Symbol is p.

Glossary

Pin Array	(*Or pin-grid array or area array*). A package with pins distributed over much or all of the bottom surface of the package.
Pinhole	A small undesired hole in an oxide, opaque region of a mask or reticle, or in a photoresist layer.
PLA	(*Programmable Logic Array*). A general-purpose logic circuit containing an array of logic gates that can be connected (programmed) to perform various functions.
Planar	Existing essentially a single plane; a process in which all PN junctions intersect the top surface of the semiconductor material, such that these intersections are permanently protected by the masking oxide, and all contacts to the device can be made to the top surface.
Planar Process	The planar process of forming integrated circuit and semiconductor components is based upon the use of a single surface for referencing each successive opeation. The planar process depends upon the repeated use of silicon dioxide (SiO_2) on the surface to control the location of impurities.
Plasma	An electrically conductive gas composed of ionized particles which are used to etch unwanted material through a chemical or physical bombardment process. Plasma etching takes place in a reactor, which may be of the barrel type or the planar type.
Plasma Etching	The utilization of RF energy added to a chemical in gas form producing a glow discharge. This glow discharge contains chemically reactive species, (atoms, radicals, ions) which react chemically with the material to be removed, whose by-products are volatile.
PLCC	(*Plastic Leaded Chip Carrier*). A leaded quad package — a replacement for the plastic DIP in surface mount applications.
PMOS	(*P-channel MOS*). A type of MOSFET using holes to conduct current in the semiconductor channel. The channel has a predominantly positive charge during conduction.
Polysilicon	Polycrystalline silicon. Sometimes call "*poly*." The form of silicon made of many small randomly-oriented crystals. Doped poly is a conductor of electricity and is often used as an alternative to metal in interconnecting devices on integrated circuits.

Glossary

Positive Resist — Photoresist that is removed in areas that were not protected from exposure by the opaque regions of a mask — while remaining in regions that were protected by the mask. The positive image of the mask remains following the development.

Predeposition — The first part of a two-part diffusion. In this part, a high concentration is diffused shallowly into the surface. This acts as a source for the second or drive-in portion of the process.

Preforms, Solder — Formed pieces of solder material for closing microcircuit packages. Additionally used in some applications to attach chip to package.

Printed Circuit Board (PCB) — A substrate on which a predetermined pattern or printed wiring and printed elements has been formed. Also called a printed wiring board (PWB).

Probing — A term used to describe testing of individual IC dice by using very fine probes to temporarily connect each to a test computer, thus verifying proper operation. A bad die will usually be marked with a spot of ink.

PROM — (*Programmable Read-Only Memory*). A read-only memory that can be programmed after manufacture by external equipment. Typically PROMs utilize fusible links that may be burned open to produce a logic bit in a specific location.

Propagation Delay — The propagation delay of a circuit is the finite period of time (delay) measured from the instant when the input signal(s) is applied until the output has reached its final value.

Proximity Aligner — An optical system that uses proximity printing to expose a wafer.

Proximity Printing — Exposure of a wafer by passing light through a mask that is very close to, but not in contact with, the photoresist-coated wafer. Collimated light is required to ensure minimum image resolution.

P-type Semiconductor — A semiconductor crystal containing a small amount of "dopant" atoms that have one less outer electron than the other atoms. Each "dopant" atom causes one unoccupied spot, called a "hole," among the electrons that are bound in their orbits. The holes are positively charged and in effect move, constituting an electric current. Boron is a commonly used P-type dopant for silicon.

Glossary

Qualification	The technique of evaluating samples to quality or approve a production lot from which they dome or the production line from which they are produced.
RAM	(*Random-Access Memory*). Stores digital information temporarily and can be changed as required. It constitutes the basic (read/write) storage element in a computer.
RCA Clean	A special wet chemical cleaning process using a combination of water, peroxide, and ammonim hydroxide followed by water, peroxide, and hydrochloric acid.
Rapid Thermal Processing (RTP)	The use of radiant light energy to heat silicon wafers very rapidly to elevated temperatures.
Reactor	A term used to describe a piece of semiconductor process equipment used for depositing various layers.
Refractory Metal	A metal that has high temperature stability. Generally refers to those metals in the Periodic Groups of IVA, VA, and VIA.
Registration Overlay	The accuracy of position of all patterns with respect to previous patterns that form other layers of a semiconductor wafer.
Rise Time	A measure of the time required for the output voltage to change from a low voltage level ("0") to a high voltage level ("1") once a level change has been started.
Resistance (R)	The difficulty in moving electrical current through a conductor to which voltage is applied. Expressed in ohms (Ω).
Resistivity (r)	A physical property of a material to resist or oppose the movement of charge through the material. Expressed in ohm-cm (Ω-cm).
Resolution	The smallest image that can be clearly discerned.
Reticle	In semiconductor parlance, a glass or quartz substrate with a 10X image of a single IC. The 10X reticle is produced by a pattern generator. The 10X pattern may be optically reduced and stepped onto a master plate, or used directly on a stepping aligner system (commonly referred to as a stepper).
Rinse	The use of D.I. water to neutralize, clean, and remove another liquid.

Glossary

ROM	(*Read-Only Memory*). Permanently stores information repeatedly used-such as tables of data, characters of electronic displays, etc. Unlike RAM, ROM cannot be altered.
Scaling	A technique of reducing the size of an existing integrated circuit die by selective shrinking of the X and/or Y dimensions.
Scanning Electron Microscope (SEM)	A specific electronic microscope that uses an electron beam to scan a specimen, under vacuum, providing a very high magnification of the surface. This magnification is generally in the range of 10x - 100,000x.
Scanning Projection Aligner	An optical system that uses a slit of light to transfer the mask image to the photoresist-coated wafer. The slit of light is scanned across the mask and is projected onto the wafer.
Schottky TTL	A form of TTL logic in which Schottky diodes are used to clamp the transistors out of saturation, effectively eliminating the storage of charge within the transistor — allowing increased switching speeds.
Scribe Line or Saw Lane	The separation between adjacent die on the wafer. This path is used as the cutting area in sawing a wafer into the individual die.
Selective Etching	Etching that is done so that certain material is dissolved, but other materials are not affected by the etchant.
Semiconductor	A material with properties of both a conductor and an insulator. Common semiconductors include silicon and germanium.
Sheet Resistance	Sheet resistance is a characteristic of material that is formed in such a way so that its resistivity is not a function of its cross-sectional structure. Sheet resistance is expressed as R_S in terms of ohms per square, where the square refers to surface area but has no absolute dimensions.
Side-brazed Ceramic Package	A multilayer ceramic dual in-line package (DIP) with external leads brazed on the sides of the package.
Silicon (Si)	The basic element used in most semiconductor devices; i.e.; diodes, transistors, and integrated circuits.
Silicon Dioxide (SiO_2)	The compound of silicon and oxygen used for insulation (isolation), passivation, or a masking layer in semiconductors.

Silicon Nitride (Si_3N_4)	The compound of silicon and nitrogen used as a masking layer during processing and as a final passivation layer at the end of the process sequence.
Silicon-Gate MOS	MOS devices with a polycrystalline silicon layer deposited to form the gate elements.
Silicides	The reaction of a metal with silicon, forming a metal-silicon compound, i.e., PtSi (platinum silicide).
Single-Crystal	A piece of material having a continuous, regular crystalline lattice structure and having no internal grain boundaries.
SIP	(Single In-line Package) with a single row of pins (in picket-fence style). Package is usually mounted in the vertical plane.
Slice	Another term for wafer.
Soft Bake	A heat treatment used after the photoresist is applied to the silicon wafer to remove the solvents from the layer.
Software	The programs and instructions for a computer.
SO/SOIC	(Small Outline, Small Outline IC). A small plastic dual in-line package, usually with "gull wing" feet, designed for surface mounting. Most versions have lead spacing compatible with that of PLCC packages. Another version, SOT (*Small Outline Transistor*) has been popular in hybrid use for several years.
SOI	(Silicon-On-Insulator). Similar to SOS, but with oxide or another insulting film isolating individual transistors.
SOJ	(Small Outline, J bend). An SOIC package with j-bend leads.
Solid-State	Refers to electronic devices, like transistors, that conduct and control the movement of electrons in solid materials (instead of in a vacuum).
SOS	(Silicon-On-Sapphire). A fast MOS technology in which silicon is epitaxially grown on a sapphire wafer, and etched away between transistors. Each device is thus isolated by air or oxide from other devices.

Glossary

Source	The working-current terminal of an FET that is a source for holes (P-channel) or free electrons (N-channel) flowing in the channel. Corresponds to emitter of bipolar transistors.
Spider Bonding	A method of connecting an integrated circuit die to its package leads. A leadframe is placed over the chip and all connections are made by just one operation of a bonding machine. Tape automated bonding (TAB) uses this approach to interconnection.
Spin or Spinning	The process used to apply the photoresist to the wafer surface.
Spreading Resistance	An evaluation technique used to measure the dopant profile of a p-n junction.
Sputtering	A method of depositing a film of material on an IC wafer. A target of the desired material is bombarded with excited ions that knock atoms from the target, which are then deposited on the wafer
SSI	(*Small-Scale Integration*). ICs containing fewer than 30 transistors.
Standard Cells	Predefined logic elements that may be selected and arranged to create a custom IC more easily than through original (custom) design.
Static RAM (SRAM)	A type of semiconductor memory in which a pair of flip-flops are cross-coupled to hold a binary state as long as power is applied.
Steam Oxide	The use of high-temperature water vapor to grow silicon dioxide. The source of water vapor can be from D.I. water or the pyrogenic reaction of hydrogen (H_2) and oxygen(O_2).
Step Coverage	The thickness of a thin film over topological changes relative to the thickness on the top surface.
Step-And-Repeat Camera	An optical system that reduces an enlarged reticle image to 1X on a master plate. After each exposure, the master plate is stepped to a new position and the exposure process is repeated.
Stepping Aligner	An optical system that projects (usually 5X or 10X) reticle images on a portion of the photoresist-coated wafer. After exposure, the wafer is stepped to a new position and the exposure process is repeated.
Stripping	The process of completely removing a coating such as photoresist.

Subcollector	See Buried Layer
Substrate	The material on which a microelectronic device is built. Such material may be active, like silicon, or passive, like alumina ceramic.
Sulfuric Acid (H_2SO_4)	A strong oxidizing acid used to clean silicon wafers and remove photoresist.
Susceptor	A plate on which semiconductor wafers are heated during thermal processes such as epitaxy and chemical vapor deposition (CVD).
TAB	(*Tape Automated Bonding*). A process utilizing metal conductors on beam tape, which are mass bonded to the IC die in a single operation.
Target	The source material for a sputtering process.
TCE (Trichloroethylene)	A solvent used for cleaning parts.
Thermal Compression Bonding	A method of wire bonding that does not use an intermediary metal or melting, but rather the plastic flow of materials resulting from the combination of heat and pressure. Also called thermocompression bonding.
Thermal Diffusion	See Diffusion
Thick-Film Process	A hybrid microelectronic process where conductors, insulators and passive components are screened from special pastes onto a substrate. This process is less expensive than the thin-film process, which uses depositions and photolithography techniques to define conductors, etc.
Thin-Film Process	The use of deposited films of conductive or insulating material, which may be patterned to form electronic components and conductors on a substrate, or used as insulation material between successive layers of components.
III-V Compound	Refers to semiconductors formed by combinations of materials in groups III and V of the periodic table of elements. Commonly refers to compounds of gallium, such as GaAs, GaInP, etc.
Threshold	The input voltage at which the output logic level changes state.

Glossary

TO Package	(*T*ransistor *O*utline) Standard mechanical packages used primarily for discrete transistors and low pin-count ICs.
Torr	The pressure exerted by 1mm of mercury. Atmospheric pressure is equal to 760mm of mercury.
Transistor	An active semiconductor device with three electrodes that may be either an amplifier or a switch.
Truth Table	All the combinations of possible states of the inputs and outputs of a circuit. It tabulates what will happen at the output for a given input combination.
TSOP	(*T*hin *S*mall *O*utline *P*ackage). A thin SOIC or SOJ package.
TTL (or T^{2}L)	(*T*ransistor-*T*ransistor *L*ogic). A bipolar technology where a transistor output is connected directly (rather than through a resistor or diode) to a transistor input of the next stage.
Tube	The quartz tube inserted into the furnace that becomes the process chamber for diffusion and oxidation.
Turn-On Time	The time required for an output to turn on (sink current, to ground output, to go to 0 volt). It is the propagation time of an appropriate input signal to cause the output to go to 0 volt.
Turn-Off Time	Same as Turn-On Time except the output stops sinking, current goes off, and/or goes to a high voltage level (logic "1").
ULSI	(*U*ltra *L*arge *S*cale *I*ntegration). ICs containing 10 million or more transistors.
Ultrasonic Bonding	A wire bonding technique that utilizes ultrasonic energy and pressure to form the bond, without heat.
Undercut	The lateral etching that occurs as the etching proceeds vertically.
Unipolar	Refers to FET devices where current passes only through one type of semiconductor material (P or N) as it flows from input to output.
Vapor Plating	A vacuum process, usually at less than 10^{-6} Torr (mm of Hg), where metal(s) are vaporized through thermal agitation, then recrystallized on cooler surfaces, generally the material to be coated. Also referred to as evaporation.

Glossary

Vapor Priming	The process for coating the wafer surface with HMDS.
VIA	A path filled with conducting material between circuit layers.
VLSI	(*Very Large Scale Integration.*) ICs that contain 100,000 or more transistors, but less than 10 million.
Voltage (V)	Electron potential in an electrical wire or circuit. Usually expressed in volts (V).
Wafer	A thin disk of semiconductor material (usually silicon) on which many separate chips can be fabricated.
Wafer Fabrication	See Fabrication
Wafer Sort	The electrical testing of each die on the wafer while still in wafer form.
Wafer Flat	A region ground into the boule (ingot) to identify crystal orientation, dopant type and mask/reticle location reference.
Wedge Bonding	See Bonding, Wedge
Welding	Joining of two or more pieces of metal by fusing them together.
Working Plates	Masks printed from master plates that are used for production exposure of wafers. As these plates are subject to wear, they must be replaced periodically.
Yellow Room	The room in which wafers coated with photoresist are exposed to ultraviolet light in an aligner. Fluorescent lights in the room have yellow filter tubes around them to block unwanted ultraviolet light from that source.
Yield	Yield is the ratio of the number of acceptable units to the maximum number possible.
ZIF Socket	(*Zero-Insertion Force* socket.) A socket in which package leads are readily accepted by the socket, then firmly connected through cam action.
ZIP	A zig-zag lead form to increase the spacing between leads.